INTERNATIONAL SCIENCE AND TECHNOLOGY

Also by Mekki Mtewa

AFRICAN ADMINISTRATION AND DEVELOPMENT POLITICS (*editor*)
THE CONSULTANT CONNEXION: Evaluation of the Federal Consulting Service
INTERNATIONAL DEVELOPMENT AND ALTERNATIVE FUTURES (*editor*)
MALAWI POLITICAL THEORY AND PUBLIC POLICY
PERSPECTIVES IN INTERNATIONAL DEVELOPMENT (*editor*)
PUBLIC POLICY AND DEVELOPMENT POLITICS: The Politics of Technical
  Expertise in Africa
SCIENCE, TECHNOLOGY AND DEVELOPMENT: Options and Policies (*editor*)

International Science and Technology

Philosophy, Theory and Policy

Edited by
Mekki Mtewa
Founder and Chief Executive Officer
Association for the Advancement of Policy, Research and
Development in the Third World

St. Martin's Press   New York

© Mekki Mtewa 1990

First published in the United States of America in 1990

ISBN 0–312–03688–4

Printed and bound in Great Britain

Library of Congress Cataloging-in-Publication Data
International science and technology: philosophy, theory, and policy
edited by Mekki Mtewa.
p. cm.
ISBN 0–312–03688–4
1. Science and state—Developing countries. 2. Technology and state—
Developing countries. I. Mtewa, Mekki.
Q127.2.I576   1990
338.9′26′091724—dc20   89–27469
                                                        CIP

To my parents for their dreams of a brighter, humane tomorrow and to Lessie Viola, for her commitment to the present

Contents

About the Authors

Eufronio Carreño Román teaches economics at Kean College, New Jersey.

Elia Chepaitis is the designer of the Elia system, an alternative to Braille. She teaches computer information management systems at Fairfield University.

William B. Crawford teaches management at Slippery Rock University, Pennsylvania.

James Grunig teaches journalism at the University of Maryland, College Park.

Larissa Grunig teaches journalism at the University of Maryland, College Park.

Peter Habermann teaches communications at Florida International University, Miami.

John James Haule teaches journalism in the School of Journalism at Mississippi State University.

Floyd Idwal John teaches business and management at Marymount University, Virginia.

Martha Tyler John is Dean of the School of Education and Human Services at Marymount University, Virginia.

Marvel A. Lang teaches urban affairs at Michigan State University, East Lansing, where he also heads the Urban Affairs Institute.

Lewis A. Mennerick teaches sociology at the University of Kansas at Lawrence.

Mekki Mtewa is the founder of the Association for the Advancement of Policy, Research and Development in the Third World and is its current Chief Executive Officer.

Mehrangiz Najafizadeh teaches sociology at Mount St Mary's College, Emmitsburgh, Maryland.

David Porter teaches political science in the Department of Political Science at Youngstown State University, Ohio.

Acknowledgments

Since founding the Association for the Advancement of Policy, Research and Development in the Third World in 1981, I have been surrounded by budding Third World professionals with exceptionally promising talent and expertise. A selection of their thinking and work is incorporated in this volume. My gratitude and thanks go to them with the hope that the dawn of the next decade unveils more professional opportunities for them as they map out new horizons of scientific discovery and technological application.

The collation of the papers appearing in this volume was first undertaken by Edward C. Pytlik at Morgantown, West Virginia, but the follow-up and completion of that process was very much facilitated by the refinement of those papers by Dr Louis Armijo. However, niether of these colleagues can be held responsible for any editorial errors of context, organization or substance in the individual contributions to this volume. The authors and I cannot escape that responsibility.

I am especially indebted and equally appreciative of the professional support of Brenda Lee Hickey who sacrificed a substantial amount of her leisure time in order to help me meet the editorial deadline for this volume. And to all our colleagues in the global community who are dedicated to the advancement of the quality of human life through science and technology – Cheers!

Foreword

Mekki Mtewa

While on a visit to the offices of the Mexican Council of Science and Technology (CONACYT) in Mexico City and again at the offices of the Indian Council of Social Science Research, New Delhi, I got to thinking about how many excellent social science and technology institutions are located in India and Mexico. Between these two countries, there are more scientists and technologists than can be found in sub-Sahara Africa and the Middle East combined. And that between Mexico and India there are not only exceptional institutions and professional talents that continue to make significant contributions in science and technology but that these two countries stood the best chances of playing leadership roles in science and technology within the developing group of countries. Within the confines of those institutions with their exceptional professional talents are thousands of individuals, each one a unique reflection of professional culture, values and politics. Every one of them was as capable of winning the Nobel Prize as the next one and each with a determination of advance the legacy of science and technology for the service of mankind.

Quite often, for those of us who spend most of our efforts and time in building communication bridges between our institutions and societies, it is important, when we talk about scientific, technological and developmental concerns first to value individual professional contributions and to make sure our institutions nurture and efficiently utilize them.

Creativity, innovation and productivity remain the most important characteristics in science, technology and development. CONACYT embraces all these. However, I see CONACYT like its counterparts in other developing countries playing also the role of a catalyst. There are millions of students entering public schools each year who, we could hope in some way, are influenced by what is happening in institutions such as this. As professionals involved in science, technology and development we should tend to be contagious, that is to say professional and individual excellence should set the standard(s) of education for all generations in developing countries.

We should not only be concerned about existing laboratories, curricula, teachers and students, but equally about professional leadership. By leadership, I mean the process of priority-setting for the 1990s and beyond, and the influences and impact which dedicated professional organizations may have. Priority-setting is not made in a value-free context; it transcends complicated boundaries. These are essential since they are structured around recognized and familiar needs. The crossing of each boundary involves some changes and flexibility in our structure of values. It is rudimentary, therefore, that decisions made without considering values and institutional capabilities would become irrelevant.

The foundations for some of our Third World development ills are clear and need continuing attention. The Third World has been plagued for years by ambitious, unattainable priorities. Declining capital and unadjustable, uncontrollable budgets, complex and inflexible structures of restrictions and an absence of incentives to encourage technological investments and growth have decelerated economic productivity, economic creativity and industrial innovation. All of these problems have a way of affecting science and technology. Recognizing the real constraints imposed by today's economic realities on science and technology, what should we do to ensure, at the minimum, the utilization of existing science and technology resources in the 1990s? Let me summarize our Association's attitude toward science, technology and the Third World future.

There is an apparent correlation between a country's emphasis on scientific achievements and adoption of technology with the overall positive direction and strength of its economic development. A good science and technology foundation could provide developing countries with some measurable ability and readiness to meet development challenges of the unforeseeable future. The predictable value of science, supported by well-endowed technologies and institutions, could guide us in our definition of optimum priorities for the future. Without this infrastructure, no one in the Third World can with certainty tell us what the alternative futures of development will be.

In contrast to the expectations and perceptions of many individuals and organizations, our Association believes that the condition of basic research in developing countries is generally good and healthy. That health is reflected in the sound exciting basic research taking place in many diverse fields. As a developing entity, the South spends more money on applied R&D in agriculture and other fields than on social science research, and generally has more scientists and engi-

neers engaged in those activities than in social science fields. As a
developing entity, the ratio of R&D to GNP in the South compares
favorably to that of medium industrialized countries in the North.
We cannot use the award of Nobel Prizes as a measure of the true
value and standard for the health of basic science in the Third World.
In the last ten years, a number of Third World scientists have won
or shared Nobel awards – and the future is likely to compensate for
their low rate of recognition in years past.

Notwithstanding the generally sound health in Third World science
and technology, maintaining and advocating it requires that we
continue to address a number of sensitive emerging problems:

1. Research related to science, technology and development must
   be clearly articulated and defended if we are to ensure at least
   some basic strength and capability.
2. Educational institutions must enshrine the values of science
   and mathematics in their curricula so that present and future
   generations are exposed to scientific and mathematical thinking.
3. Priority-setting should take advantage of the increasing sophistication and available technologies within our centers of learning
   and affiliate organizations.
4. Comparable emphasis should be placed on basic as well as social
   science research so that a constant infusion of interrelated ideas,
   talents and perspectives is assured.

In addressing these issues, the Association for the Advancement
of Policy, Research and Development in the Third World recognizes
with considerable satisfaction the readiness of Third World professionals to excel. There are a number of good reasons why we feel
professional communication between Third World professionals is
desirable even though we cannot expect it to pre-empt political
protocol in all fields. The realization that we may not be the first to
cross the spectrum of protocol is not simply a function of our own
maturity as Third World professionals. Rather, we prefer to look at
our common efforts as a prerequisite to the future celebration of
science, technology and development as well as professional freedom
and scholarly responsibility. As a group of Third World professionals
we continue to ask about the equitable distribution of resources
between science, technology and development and also about the
relative effect of these apportionments on social and cultural activities.
Our perception is that priority-setting can do a great deal in determining the ranking of these apportionments. We have to be rather

courageous in facing the future and remember, too, that science and technology should not lag behind in the way they have in the previous three-and-a-half decades. The scientific and technological professionals in the Third World should participate in priority-setting exercises before mandatory budgeting allocation decisions are made. This Association's view is that unconsultative budgetary decisions everywhere are subject to inherent cynicism. Science and technology suffers because it is, quite often, not clearly quantifiable and rationalized with the help of cost-benefit ratios.

To summarize, it seems evident that tomorrow's development issues will be different from those of today. A changing view of priority-setting in national development calls for changes in professional roles and policy perceptions. Science, technology and development priorities of the future should not be a regurgitation of old formulae and indefensible growth ratios and strategies. It is in this context of shifting professional responsibilities that we invite and encourage our colleagues in developing countries to join together and embark upon the search for an enlightened scientific and technological agenda through an elaborative pursuit and commitment to excellence in science and technology in development, and forever repudiate developmental pursuits devoid of these foundations! I am proud to be the editor of this definitive book.

Section I
Science and Technology for Development

1 Science and Technology for Development: Theory and Practice

Mekki Mtewa

Since the formulation of the development thesis in the 1950s, the literature of institutional evolution has been erratic and diverse. The concern of scholars continues to be with the evolution of institutions in the developing world, and not with the adequacy of the evolutionary theories themselves. In this literary preoccupation almost every theoretical proposition, including those of the previous evolutionary utility theories, seems highly debatable.

The participants in this debate are many, and the interdisciplinary distances between them continue to widen.[1] The debate on development economics and policy takes place simultaneously in both famous and obscure forums. Highly perceptive articles appear in little-known journals. Because of restrictions over policy access and input, these pieces may get read, but not by the policy-makers who are desperate for the solution. The vicious circle of the poverty of development policy is, in part, caused by the unfair advantage some disciplines have over the dispensation of knowledge within the social science.[2] In the literature on institutional development there is also the problem of professional style, and preoccupation with models and theoretical abstractions. All these have their usefulness in scholarship. However, preoccupation with these to the neglect of effective communication for practical policy creates a lasting gap between theory and practice. This gap contributes to a crisis in effective communication between the professed policy experts and the policy practitioners. George Stigler observed that most economic literature on development policy written by economists is written for other economists, and the models used there are either in repudiation or support of other models which only economists of similar persuasion could comprehend.

Professor Rostow advanced the popular notion that development can be planned in stages.[3] According to this doctrine, development is a linear path along which all countries inevitably travel. Western countries have, at various times, each at their own speed, passed the stage of 'take-off'. The developing countries are now said to be following this pattern. According to this view, development is a process of leadership by objectives and the securing of social goals through sophisticated tools of planning.

However, Rostow's notion of linear development planning has been subject to some criticism. From a political perspective, Rostow's linear view ruled out the options of different perceptions of development and the variable tools of planning appropriate to each of these. For preferring one particular path over another, he has also been criticized for being excessively deterministic. Determinism, when it is adopted as a preferred planning approach, narrows the planning options to only one. This myopic view clearly does not reflect the world's diversity.

There is a second view which has gained currency following the spread of disenchantment with planning in stages and the fundamental deficiencies of such a view in practical policy. 'Government by objectives' looks at the economic disparities that exist between countries. According to this view, planning is not a transitional phenomenon to be ended after the 'take-off' stage, but a permanent qualitative feature in the development experiences of non-Western countries.[4]

Professor Rostow, representing the first planning view, and André Frank the second, are the two poles of the development planning argument rather than the policy models. Albert Hirschman has developed an integrative approach from these two views.[5] In this approach, Hirschman raises fundamental questions as to whether everything governments planned for could be achieved, even with the help of free trade policies, established competitive markets and liberal monetary policies. These issues, theoretical though they seem, are the mainstream, not the periphery, of the formulation of a comprehensive development policy. Recent proponents of this comprehensive development view for the Third World now include the World Bank and a majority of its financial subsidiaries.[6]

Richard Rose has three suggestions that are appropriate to our understanding of the dilemmas of the development-oriented politician.[7] The first is that politicians, proceeding from a purely political perspective, do not assume policy positions with a *comprehensive* development policy. They are vaguely aware what elements and components of a development policy are appropriate for the maximization of development objectives. Secondly, the politician appoints his Cabinet ministers and exercises control over them without much appreciation of organizational efficiency. The processes of being a politician and at the same time an organizational student in the face of political uncertainty produce a contradiction in the terms of policy credibility and expediency. The third of Rose's suggestions is that, while politicians assume their duties deficient in technical expertise, they lack appreciation of those professional staff members who are more technically gifted than they. As a policy-making arrangement, in fact, the relationship between politicians and professionals is built on mutual contempt. The contradictions are many and, indeed, the risk to those professionals who blindly adhere to the principles of their disciplines is clear.

In making development policy decisions, development leaders recognize that the manner of determining development preferences is significantly different from the postulations of theory. Therefore, not all development theory affects the shapes of preference curves. In the Third World, a development leader operates from no assumption of a conflict of interest. His office and person, both in law and in fact, supposedly embody and reconcile these.[8] The remainder of this chapter first reviews the constituents of development, then the grounding of science and technology policy, and finally the introductive models and theory of international science and technology preferred by a sample of Third World governments.

DEVELOPMENT POLICY

The argument proceeds from the premise that development-oriented research has a contribution to make to the planning activities of development-oriented governments. The first task here is to define goals, objectives and needs and give examples of what, by reference, each operational term means. The second task is to consider how goals, objectives and needs relate to and differ from each other.

Goals are what a development-oriented economy and its organizations might try to accomplish. *Objectives* are those specific aims of the developing government and its agencies toward which, in order to accomplish the government's declared goals, public resources are committed. Like goals, objectives describe what an organization is striving for, but these differ from goals by being more specific. Objectives, from a budgeting point of view, appear in the form of budget allocations, per project or program, of a specific sum of money to be spent over the budget period. One may assume, from a budget point of view, that meeting the objectives will contribute to accomplishing the goals. This assumption takes into consideration the requirements that (1) an objective should be measurable; (2) it should indicate a target date for its accomplishment; and (3) it should specify the degree of achievement desired.

Having a clear development policy that makes it possible to select goals from a multiplicity of development issues leads to a more efficient allocation of scarce resources and provides a basis for setting priorities. A developing government with coherent goals that appear within a comprehensive development policy provides itself with a good point of reference. It is able to measure, strengthen and/or eliminate development issue for which its available organizational structures are not prepared and ready.

With coherent goals, it is easier to evaluate a government's performance and allocative efficiency. A starting point of performance evaluation is usually the determining of what a government wished to accomplish and the comparing of this with what it actually did accomplish. Against this framework, however, performance evaluation also looks at the opportunity cost element; that is, what it actually would have accomplished if all cost-benefit and management considerations were taken into account.

Apart from the political implications of goal-setting in a development-oriented economy, responsibility for the failure to meet a government's goals must be jointly shared among all the sectors of the economy. Goals that deal with public development sectors should involve all sectors of the public economy. Sectoral goals, however, are individual since they involve distinguishable interests within the economy. In the aggregate, however, sectoral goals have a direct impact on the government's development productivity.

An analysis of the intricate relationship that exists between ideological commitment and the outcome of the development effort reveals that these two elements of human concern cannot be happily separated. From a political point of view, political or ideological declarations are good indicators of the policy intents of a leadership. In administrative terms, institutions may be better off by conforming to and incorporating into their behavior the requirements and demands of these ideologies. The policy formulation and implementation process does, in such an event, incorporate an ideological filtering process, a process which those in policy positions might use to select those segments of the ideology that best complement their policy objectives.

To minimize problems of understanding, a development-oriented government might advance the view that the search for development objectives does not end at the State House. The Office of the President or that of the Planning Division is the legitimate initiator of development programs. The clarification of these, however, involves many sectors, both internal and external.

Development strategies are difficult to assess since development-oriented governments have a tendency to describe strategies before they have thought through the needs and goals and analyzed their organizational or technical capabilities. The advantage of an active Presidential development strategy is that it might focus on what she/he knows can best be accomplished rather than examine what must be done to accomplish those development objectives. However a Presidential strategy might be misguided and its goals might be in conflict with clearly established options.

While not all decision points can be foreseen, a thorough examination of development strategies, in the light of diverse technical resources and obstacles, can expose strengths and/or weaknesses that might be useful in the evaluation process. An analysis of evaluation technicalities demonstrates that it is desirable in development-oriented policy research to point out the appropriate role that this process might play.

Evaluation is a fourfold process of periodic (1) assessment of programs that are meant to achieve the highest possible effectiveness; (2) monitoring and review of development programs that are meant to conform to funding requirement; (3) identification and selection of programs that require the use of multiple tools of analysis; and (4) definition, refinement and application of different stimulative and regulative policies conducive to the whole development effort. Development evaluation serves an additional use in that it allows the establishment of a comprehensively formulated development philosophy. This philosophic component is what appeals to the various sectors and motivates each of these to contribute equitably to the realization of the goals.

Development evaluation should not be thought of as a tool to prove how poorly a government or its development policies have performed. Development evaluation is a tool for improving government or program efficiency or effectiveness. Adopting this positive attitude, development evaluation becomes a necessary part of any effective development planning and budget allocation policy.

Needs are the ends which a development-oriented government pursues. While goals and objectives are related to needs, the process of classifying development needs either takes an imperative or an optional form. Imperative needs are those with a government would want to initiate because they are pressing. The prioritization of imperative needs may either follow an objective or a subjective view since the assignment of macro or micro values to some purely human or social needs may not be politically prudent. Richard Rose explains the problem this way:

> A prudent incumbent . . . would usually approach imperative issues by taking a non-directive initiative, and then, after events had unfolded and support for policy options became clearer, respond with a specific statement of objectives appropriate to the problem at hand.[9]

Needs dictate strategies. A development-oriented government proceeds from the view that all human needs can be met since the lessons of Western technology share this optimism. A need might, at its minimal level of clarity, guide policy and how to get there from here. A politician may find difficulty in believing policy analysts who tell him so, although common sense simplifies the movement to there from here. There are numerous policy options, costs and considerations of efficiency that should enter into the final decision. In the development effort, development evaluation should form an integral part of a development goal-identification, plan-implementation, and program-review process.

Development needs, plans, strategies and organizations change. A good development policy permits self-evaluation to assess its own performance and the incorporation of the result of that performance into the revised development policy. By this definition, development evaluation is the act, or result, of comparing what is actually happening to a development policy with what the government initially intended or planned to accomplish within the development period. Using this definition, the performance, or effect, of a development policy can easily be modified with respect to where the development philosophy and interests of the sectors are, and how these approximate the comprehensive development policies, goals and objectives of the nation as a whole. This performance evaluation can be done in terms of sectoral efficiency, a process by which the sectors are to articulate their needs and formulate their strategies while the national policies guide their sectoral options. The process of prescribing uniform strategies lacks the appreciation of flexibility and innovation in economics and imposes 'no-growth' options on growth-oriented sectors. Evaluation may reveal how different strategic options are related to 'development with growth' and 'development without growth'. Pickett's criticism of Ghana's first Seven Year Development Plan focused on the paradoxical linkage between no-growth development and growth-oriented development.[10] A government might realize too late, as did President Nkrumah, that his development goals and objectives had side-tracked developmental rationality.

Measurable result is the end result of the performance evaluation process. Measurable result is purely a planning and budgetary tool. It helps planners to determine whether the objectives of a given project have been accomplished. For purposes of prioritization, policy-makers can intelligently compare the impact of identical programs under similar circumstances and make budget allocation decisions based on this determination of measurable program impact.

Setting clear, measurable objectives appears to be a desirable prerequisite in evaluating the outcome of development efforts. Development expectations, at least if they are not to remain merely a vision or in the realm of fantasy, should be brought into the realm of the possible. It is reasonable to expect that developing governments should decide what should be included in a development policy and, during the policy implementation and evaluation processes, specify policy responsibilities and who is accountable to the central policy agencies for getting these tasks completed.

Action research strategies might be useful to development-oriented governments in the process of identifying who has competency to conduct what policy function or task within the program and how this structure of authority might contribute to the overall administrative efficiency and coordination requirements of the overall development effort. Here, it is necessary to recognize that a number of development programs intersect and involve identical agencies or ministries. Masakhalia provides an illustrative example of this problem in Kenya:

> In the past, projects were held up for two main reasons: (i) they were insufficiently prepared in anticipation of choices and decisions to be made at subsequent steps and stages, and (ii) practical problems of implementation, including co-ordination among associated agencies.[11]

The management process for program coordination prescribes, however, that in such an event the sharing of management responsibilities over identical programs should be legislatively provided. The extent of this provision, however, depends on the active role the legislature plays in the overall development effort.

TECHNICAL UTILIZATION

The professional has become a symbol of expert advice, and a great deal of public attention is devoted to what he does, how he does it, and the uses to which the products of his professional work are put. However, the professional is limited in what she/he does by direct reference to the terms of his employment. She/he has little choice, except through setting a range of probable options for his employers or sponsors, in what happens to his professional products.

In reality, however, the professional in a science and technology oriented economy should have more functional and structural support than constraints that impinge upon his policy input. Relating this to planning, Kamarck describes an expert or team of experts arriving to prepare a development plan.[12] They set to work to construct or fill in a model of the economy, build up intersectoral input-output tables, construct a system of equations, and present an internally consistent overall development plan complete with capital/output ratios; marginal savings coefficients; import, consumption and production functions; investment; savings; and import gaps. A development plan of this nature, according to Forrest, is only vaguely operational in the sense that decisions to comply with the identified requirements are made by the developing government without the help of the experts who formulated the plan in the first place.[13] Technical complimentality is needed in order for this kind of planning to succeed.

Developing governments, according to our understanding, have unreasonable expectations of the help of foreign technical professionals to the neglect of their indigenous professionals. Yet it is this indigenous professional component that must effectively administer the sophisticatedly formulated development plans (see Chapter 5). This neglect creates a professional imbalance, in Hirschman's view, between the volumes of formulated development plans and those that are implementable.[14] Lewis provides a further critique of this problem:[15]

Mathematical models are needed by countries where the growth of internal demand is the engine of development, since full identification of the possibilities then requires demand projections and input-output analysis of inter-industry transactions. In Africa development planning is primarily an exercise in detecting new opportunities; its tool is not mathematics, but lavish expenditure on surveys and research.

Professional utilization should, by definition, imply an understanding of the principles which guide the thinking of its members. Knowledge is the main commodity of this group of individuals. Those who possess it and create it may be accorded thereby an influence on policy decisions that fall in their areas of expertise. Experts may appeal to their expertise to justify policy positions if given an opportunity to do so. The sponsors or employers of such experts, however, must finally make the choices among these divergent policy justifications, with the realization that professional disciplines are diffuse and complex. The techniques emphasized by each professional group also vary.

Blau proposes a definition of professional orientation. This includes 'a common identification with professional values and norms, which makes the process of attaining professional objectives a source of satisfaction'.[16] Using this definition and that of Thompson, we find that a professional is an individual who not only has internalized values of autonomy, success, standards of measurements and a sense of doing professional things, but also regards highly opportunities to contribute to problem-solving and appreciates institutions that protect him from violation of his fundamental needs.[17]

Professional utilization in a development-oriented economy implies, to a greater extent, political relationships between professionals, practitioners and politicians that deviate from a desirable professional, interdisciplinary model. Professionalization in a technical society, according to Redford, is a necessary complement to official roles.[18] According to this view, the interdisciplinary dependence is based on the premise that official roles will not produce effective results unless the action is taken by men who themselves possess, or are assisted by persons who possess, specialized qualifications and attitudes, although the attitudes and opinions of these professionals might be different from the interests, attitudes and opinions of the officials.[19]

In a development-oriented economy, due to the vagueness of

political goals and objectives, it is the professional who must capably articulate, implement and attempt to educate both the administrator and the politician on the alternative options available to the government and the constraints of each development policy model. The stakes of professional risk and uncertainty are high in this relationship since they are based on meeting the unreasonable development expectations of politicians whose perception of professionals is that they possess extraordinary powers to predict and to accomplish, in concrete terms, what they profess on paper.

The seconded professionals – a class to which belong all those overseas technicians, engineers, planners and economists who come from foreign civil services – are safe from political repercussions since the extent of their accountability to their policy recommendations is limited by the terms of their secondment. The burden of proof falls heavily on those local professionals who find themselves in a position of policy significance. The local professional, unlike the seconded one, has the insurmountable task of clarifying to the administrators or, if he is one of them, to his superiors who, in this case, are politicians, the variances between the development projections and the actual. He must further justify why he should continue to occupy his position. Public Service Commissions provide little help in the form of coherent rules and regulations that might be used to guide the relationships between these two sets of individuals. In the absence of these procedural clarifications, politicans exercise extraordinary prerogatives in decisions to dismiss, retire or politically detain local professionals or administrators whose interpretations differ from theirs.[20] Professional utilization in this development environment is difficult to assess.

Using the definition of professional values previously provided, professional utilization directly relates to the professional recruitment practices of developing government, the relative uses to which they put overseas and local professionals, levels of professional protection the government gives to each of these, and the effect relative rates of professional remuneration have on the overall 'brain drain'.

Politics has three effects on professional utilization. The first is its distributive effect. This occurs when a government wishes to accomplish an employment objective to the exclusion of professional autonomy. The brain drain, as a product function of this policy neglect, might be a deliberately adopted political choice which governments may occasionally use. By their political choice, they have a clear understanding of the impact of their policy and the desirable alternative recruitment possibilities at the government's disposal.

The second political effect is of a redistributive nature. Redistributive effect bears some resemblance to the earlier Africanization or localization policies of the newly independent countries. The emphasis during this process was to recruit as broadly as possible and to fill professional and executive vacancies with relatively capable indigenous personnel. The overall effect of redistributive hiring was not felt until a decade later. The local professional personnel in Africa were subjected to two pressures. The first was that they had to meet their professional expectations of their governments relative to those of overseas staff. The second was that they had to measure up to higher and unreasonable standards set by the government as recruiters themselves. The governments, on the other hand, had to justify their individual localization policies to external funding sources and use these performance records in future justifications for such policies.

The interaction of the distributive and the redistributive policies produces an assertive, regulatory policy. The third policy effect refers to the assertive regulatory policies of the governments on the professional standards and performance levels of its professional staff. Such policies narrow professional standards and performance expectations to precisely those which the governments set for themselves. In a regulatory system it is forbidden to engage in practices or professional activities that, in the government's judgment, are of a speculative nature and may be contrary to its development expectations of what the professional component should look like.[21]

The lamentations of developing governments on the shortage of professional personnel may be assessed against this analysis. In the late 1950s and early 1960s most developing governments inherited administrative systems whose requirements for efficiency were high. The fundamental principle of this efficiency was the provision of an impartial, dedicated civil service; politics had less bearing on its consideration. Unless an appropriate dichotomy between politics and administration can clearly be made, the use of double standards in measuring professional productivity and accountability will increasingly widen the credibility gap between seconded and local personnel and produce a crisis of confidence between the leadership, the management teams, and all those involved in the development management effort.

SCIENCE UTILIZATION

Science utilization is a goal-setting, program-securing, and problem-solving process. It demands the same rational considerations as are involved in the setting of development objectives. Effective science utilization requires a clear perception of its techniques and institutional requirements and a clear estimation of its contribution to development efficiency. The responsibilities over implementation fall on institutions that normally possess the required technical expertise. However, social accounting for the overall effects of science implementation is a responsibility shared between the institutions, their personnel and the moral/political leadership.

Recommendations for the use of science in problem-solving requires an assessment of the adequacy of the evidence science provides. It also requires the incorporation of this evidence into the existing institutions. The assessment of evidence must look at the available technical expertise of the staff. Any decisions that are made to adopt scientific recommendations should be guided by opportunity-cost considerations. Opportunity-cost decisions involve the modification of non-feasible recommendations weighed against one another. The opportunity cost of policy alternative X is the amount of benefit that must be sacrificed in order to produce more of policy alternative Y. On the basis of cost-benefit estimates alone, government might shift the financial allocations for project X administered by institution A to institution B, without regard to the social, political and efficiency ramifications of such a decision. The decision that has to be made is should a reduction of project funding or institutional staff be made, moving these resources to project Y under institution B? In this case, the marginal opportunity cost of project X must be less in order for the reduction to be made. In a perfectly functioning system of science utilization, the political cause and effect of this resource allocation decision would be equal or greater than that of the forgone cost. Hence, where science utilization costs differ, as in the developing economies, the costs of science production and implementation with be minimal and inconsequential to the overall development experience.

What then must happen to the quality and quantity of the scientific information that must go into the planning of development policy? This predicament becomes evident in the consideration of issues of science policy formation and the requirements and uses of technology-oriented information. Technology-oriented information is that which critically assesses human needs and their relation to social, economic, political and scientific policies. In practical terms, however, information is useful in establishing an equilibrium of policy expectations and objectives among the variable segments of society, if the postulates of such policy are clean and coherent in scope, and adequate in content.

Proceeding from postulates of limited resources, variable institutional capabilities and desirability of alternative development options, research utilization may be defined as the effectiveness with which developing governments utilize available scientific evidence when implementing development policies. Research utilization becomes especially critical where the availability of research staff is limited or of a short-term nature.

Limited resources here refer to the distribution of available funds and facilities and personnel with the desirable levels of technical or professional competency in order to implement or incorporate effectively new recommendations into existing procedural/operating documents. There is also the element of technical uncertainty about the kinds of scientific recommendations to adopt and the probable effect these might have on the operating agencies available to the beneficiary government.

Desirability of alternative utilization options should take into consideration the institutional limitations of the user agencies and their resources, both technical and financial. Alternative utilization options may include three basic considerations. The first is to know those research or policy components that are of a priority nature. This institutional choice might be influenced by the urgency of the research findings for policy operations and the need, on the part of the developing government, to adopt it uniformly. This process is usually followed by a period of trial and error before central decisions can be made on the overall effectiveness of the utilization task and which elements of that policy to emphasize.

The second consideration is that users must know how far to go with, or apply, any new scientific recommendations. These are decisions that have to be made bearing in mind the overall policy objectives of the government and its operating institutions. Evidence of program effectiveness obtained in one country might not be entirely usable in another. Decisions to adopt parts of evidence should be made in the light of policy flexibility. Any measure of partial program effectiveness, before the adoption of entire recommendations, should be guided by considerations of marginal opportunity costs. Decisions such as these should be related to the policy options that are available to the government itself and its operations.

Thirdly, the options of where and when to test any new policy recommendations should be sifted through gradually and in stages. Decisions such as these take time and should also take into consideration the different capabilities of different institutions, and cost-effectiveness even in the same institution. The Ministry of Planning and Development, for example, may be in a position to use and effectively incorporate into its program evaluation procedures macro and micro analytic methods. The Ministry of Agriculture, although it is directly responsible for the management of these tools, might question their usefulness, and express reluctance to use these new methods which require a complete re-learning process for which there is no time allocated. Also, the Ministry of Agriculture might find itself in the predicament of policy conflict over what the central government originally set for it to accomplish, for which its own tools and methods are quite sufficient, as opposed to the methods and tools other scientifically inclined ministries would want it to use. The final determination is one which the central government must make, but only after the merits and requirements of utilization are assessed and after it has been concluded whether the need for homogeneity really exists.[22]

Institutions are by-products of legislation, government policies and administrative practices. Institutional effectiveness and innovation may, by legal design, be conditioned by the extent to which enabling statutes are either (1) imperative, or (2) discretionary.[23]

Imperative authorizations involve those which the institution was originally established by law to be responsible for. The legislation is specific, too, on the processes of decision-making which are necessary for it to function. There is also reference to the extent to which its institutional policies are binding and what may be done to enforce its authority.

Discretionary authorizations derive from the interrelationships between institutions and their policies. By implication, discretionary authority refers to how convention, administrative practice and behavior assign responsibilities amongst institutions. Ordinarily, a number of institutions may each be affected to various degrees by a particular problem. The institution that may be most affected by the issue might, however, not possess the competency to resolve it effectively. By convention the institution that is most competent to handle the problem is assured an active role in doing so. Discretionary authorizations do, incidentally, reflect established government policies and the operating behavior of related institutions. The process of homogenizing research utilization inevitably encounters problems of this nature.[24]

THEORETICAL ASPECTS OF SCIENCE AND TECHNOLOGY POLICY

There is a preliminary question which must be asked: how do we know the priorities for science policy when everything is a priority? The STD (Science and Technology for Development) economist believes that for any given priority there are set priorities. Set priorities can be used to define the same phenomena. As such, STD economics and its empirical methods rely on theories of value correspondence because established priorities are based on observable frequencies of goals, needs and objectives.[25]

Economics emphasizes that the answers to STD policy questions are secondary to the foundation upon which the methodology is based. The priorities derived by STD are therefore affected and dominated by the methodology of economics. What economics ultimately conveys as priorities are but partial answers to political questions since economic understanding of STD priorities is very much grounded in partial knowledge.

In discussing STD policy and its economics therefore, one must accept the idea of a question with no answer. Numerical symbols are not answers.

STD policy might be defined as the authoritative allocation of development values for as many developing sectors as possible, with the intention to achieve parity in growth and distributive equity in the developing economy. The government alone can authoritatively allocate these development values.

STD policy confines the usage of its activities to broad or general questions and uses other terms for detailed choices made within the framework of its development values. Thus, STD policy consists of a general frame of rules, and while the precise boundary between development policy and non-STD oriented policy is nearly always debatable, the distinction calls for a clear delineation of STD meaning.

The meaning of STD policy is rooted in the assumption that it is goal-oriented or purposive. STD policy, according to this view, means those activities that are calculated to achieve the specific or general development goals or purposes. According to this view, STD policy might involve a projected program of goal values and practices for the institutions themselves. It is a general assumption that if a government chooses to do something, it must have a goal, objective or purpose. Realistically, the notion of policy in this case must include all forms of government activity. In similar terms, government inaction must also be considered as a form of policy.

STD policy is not a new concern of social science: the writings of Plato and Aristotle are preoccupied with these concerns. Yet the major focus of STD in social science has never really been on policies themselves, but rather on the institutions and structures of government, and on the political behaviors and processes associated with policy-making.

Today, the focus of social science is shifting to STD policy – to the description and explanation of the causes and consequences of technology and science. This involves a description of the content of technological growth, an assessment of this growth on the developing societies, and an analysis of the effect this technological process has on developing institutional arrangements.

Why should social science devote greater attention to the study of STD policy? First, STD policy can be studied for purely scientific reasons: to gain an understanding of the causes and consequences of policy-related decisions with the intention to improve knowledge about society. Second, STD policy can be viewed as a dependent variable in the development matrix, and we can ask what organizational forces and professional characteristics operate to shape the content of the development policy. Third, STD policy can be viewed as an independent variable in the authority matrix and we can ask what impact STD policy has on this authority system.

STD policy can also be studied for professional reasons: an understanding of the scientific intricacies of STD policy permits us to apply social science knowledge to the solution of practical problems. Presumably, if we know something about the dialectics shaping the form of STD science, and the relative effects these dialectics may have on the outcome of STD policy, then social science, as a process, would be in a much better position to anticipate the utility of the whole STD effort.

Finally, STD policy can be studied for political purposes: to ensure that governments and agencies make appropriate policy decisions to achieve the most estimable goals. Third World scientists cannot be silent in the face of great social and political crises, and social scientists have a professional obligation to advance specific public policies.

Whether one chooses to study STD policy for scientific, professional or political reasons, it is important to distinguish its purely policy analysis function from policy advocacy. Explaining the causes and consequences of various policies is not equivalent to prescribing what available ranges of policies governments ought to pursue. Learning why governments do what they do and what the consequences of their policy choices are is not the same as saying what governments *ought* to do, or bringing about changes in what they do.

Development advocacy requires the skills of rhetoric, persuasion, organization and clarity. Development analysis encourages scholars and students to attack critical policy issues with the tools of systematic inquiry. In short, development analysis might be labeled the 'thinking man's response' to demands that social science become more relevant to the problems of developing societies.

Specifically, development analysis may involve: (1) a primary concern with explanation rather than prescription. There is an implicit judgment that understanding is a prerequisite to prescription, and that understanding is best achieved through careful analysis rather than rhetoric or polemics; (2) a rigorous search for the causes and consequences of public policies. This search involves the use of scientific standards of inference; (3) an effort to develop and test general propositions about the causes and consequences of social science knowledge or policy, and to accumulate reliable research findings of general relevance. The object is to develop general theories about STD policy which are scientifically consistent and reliable, and which apply to different governmental agencies and different policy environments.

Development analysis offers the serious student of development an approach to political development which is both scientific and relevant. The insistence on explanation as a prerequisite to prescription, the use of scientific standards of inference, and the search for reliability and generality of knowledge can hardly be judged 'irrelevant' when these ideas are applied to important policy development questions.

It is easy to exaggerate the importance, both for good and for ill, of the STD policies of governments. It is not clear that government STD policies, however ingenious, could cure all or even most of developing societies' ills. The analytic component of STD policy cannot cure all inconsistencies in a particular government policy when there is no general agreement on the extent of the theoretical and the practical policy gap such a science policy should bridge.

Perhaps the most serious reservation about the usefulness of STD policies is the fact that problems of development choice are so complex that politicians, practitioners and social scientists may not be able to make accurate predictions about the impact of selected policies unless these have been evidenced elsewhere. But whatever the preferred mode of analysis, it can be contended that the theories to guide such thinking would follow the path outlined below.

Six theoretical modes are examined in the following pages. Each has its own merit in the study and understanding of the choice of STD policy made by governments.

1. Systems theory

STD policy may be viewed within systems theory. Systems theory, according to Easton,[26] is a structure of the political system on its input and output processes and the decisions that are brought to bear upon it. The political system is that group of interrelated structures and processes which functions authoritatively to allocate policy values for a given society. Outputs of the political system are those authoritative allocations of the system, and these allocations in a developing context may constitute development policy.

Easton's view of the concept of system implies an identifiable set of institutions and activities that function to transform demands (inputs) into authoritative decisions (outputs) requiring the support of society. The concept of system implies that elements of the system are interrelated, that the system can effectively respond to constraints in its environment, and that it will do so in order to preserve itself. The use of science and technology by the system may facilitate and help refine the quality of these inputs (demands) into translatable policy and methods for meeting these demands.

2. Elite theory

STD policy may also be viewed as the preferences and values of a governing élite. Thrasymachus, in Plato's *Republic*, argued, as did Aristotle in *Politics*, that justice is in the interest of the stronger. Elite theory would view STD policy as a product of the professionally stronger and more influential social forces in society. It can be argued, therefore, using the politics of knowledge and the axiom that knowledge is power, that the future of development and Third World STD may be shaped by élites.

What are the implications of élite theory on STD policy? First of all, élitism implies that public policy does not reflect the demands of the people so much as it does the interests and values of élites. Change and innovation in STD policy may assault the aristocracy of birth and an oligarchy of wealth. Democratic theory begins with an assault upon the élitism of hereditary status, and privilege. When STD policy is defined in élitist purposive terms, then its public policy orientation will tend toward these élitist tendencies for the benefit of the few.

An élitist STD policy might suggest that élites share in a consensus about fundamental norms underlying the developing system and on the basic rules of the STD game. The stability of STD policy, and even its maintenance, may depend upon the continuation of this élite consensus.

3. Group theory

Group theory, like political pluralism, begins with the proposition that interaction among groups is the central fact of politics. Individuals with common interests come together formally or informally to press their demands upon government. And if those professionals interested in STD succeed in this game of expectations then they may be the ones to benefit from such an outcome.

The task of STD policy under group theory is to elucidate the issues as clearly as possible. In doing this, STD policy could minimize conflict (1) by establishing the rules of the game in the selection of issues; (2) by arranging policy compromises and balancing policy interests; (3) by enacting policy options in the form of strategies; and (4) by holding groups accountable to, and responsible for, their policy preferences. According to this view, STD policy is the equilibrium reached in the decision-making process. This equilibrium is determined by the relative influence of the political group involved. Changes in their relative influences could result in some changes in STD policy; that is, STD policy could move in the direction desired by the predominant group gaining in influence, and away from the desires of groups losing influence.

4. Rational theory

The democratic character of STD activities is at the heart of rational theory. Rational theory assumes a state of mind that correctly perceives cost-benefit analyses of each policy. In order to maximize net value achievement, rational theory suggests that all concerns and values are, or potentially can be, known.

To select a rational STD policy, policy-makers must: (1) know all the society's value preferences and their relative weights; (2) know all of the policy alternatives available; (3) know all of the consequences of each policy alternative; (4) calculate the ratio of achieved to sacrificed societal values for each policy alternative; and (5) select the most efficient policy alternative.[27]

This rationality assumes that the value preferences of society as a whole can be known and weighted. It is not enough to know and weigh the values of some groups and not others. Rational STD policy-making does, therefore, require unrestricted access to sources of information from which alternative policies may be constructed. Rational STD policy increases the predictive capacity of policy-makers to foresee accurately the consequences of alternative policies, and the intelligence to calculate correctly the ratio of costs to benefits. Rational STD policy-making requires, too, a decision-making system and inclination which facilitates rationality in the complex policy formation process.

5. Developmental theory

Developmental theory may view STD policy as related directly to the levels of political sophistication and confidence the STD policy-makers may acquire in the effective use of their tools and policies. Developmental theory in STD policy, in part, provides a critique on the traditional models of decision-making. According to this critique, policy-makers do not suddenly become experts in policy-making. They have to learn its techniques in order to make judgments of its value.

A developmental view of STD policy may be constrained by time, intelligence and cost. These factors may prevent policy-makers from identifying the full range of policy alternatives and their consequences. Constraints imposed by politics could also prevent the establishment of clear-cut societal goals, and the accurate calculation of cost-benefit ratios. Developmental theory may yet come to recognize the impractical nature of "rational-comprehensive" STD policy-making at an earlier stage in a country's political growth, and prescribe a more conservative process of decision-making and STD planning.

Developmentalism may be conservative in as much as existing policies that have proved their worth tend to be preferred. The policy efforts of the government may also be concentrated on strengthening them. By adopting this conservative attitude, developing policy-makers generally accept the legitimacy of established policies and tacitly agree to continue them. It is also safe to be conservative since new policies are bound to be beset with new uncertainties for which policy-makers are not prepared.

It is accepted wisdom, according to Dye,[28] that policies tend to persist over time regardless of their utility, that they develop routines which are difficult to alter, and that individuals develop a personal stake in the continuation of policies and practices which make cost-effective changes very difficult.

6. Language theory

Language theory evaluates and pinpoints biases, ideology and standards of criticism implicit or explicit in a given STD argument. Criticisms of STD theory would conform to a paradigmatic debate that would permit simultaneous evaluation of its theory and the predominant theory of development politics it supports. Aristotle in *Politics* and *Poetics* developed a paradigm for explaining the relationship between politics and the use of language. His paradigm transcends both language and politics as formulated by Kuhn in his structure of scientific knowledge. But both men can help to better structure a justification for science and technology as against other societal goals.

A philosophically sound use of language theory must account for the relationship between language and knowledge of STD and development policy in particular, as well as the relationship between politics and its personality symbols.

Willard Quine constructs a sound empirical paradigm for this purpose. In it, Quine views all science as the totality of our so-called knowledge or beliefs, as a field touching upon experience only at its extreme edges. All knowledge in the field is logically interconnected; the field is a logical system. 'If this view is right,' says Quine, 'it is misleading to speak of the empirical content of an individual scientific statement . . . especially if it is a statement at all remote from the experimental periphery of the field.'[29] If Quine is correct in this view, then no aspect of STD science can claim to have a handle on 'policy' without incorporating into its empiricism an unempirical form of a language of STD and politics. The understanding of STD, including its policy, therefore, is politically bound.

Using language theory, we seem to be faced with a dichotomy; on one hand, with Quine, we have no method for determining the political significance of STD policy statements; on the other hand, with Aristotle, we have no sound basis for natural science. However, Aristotle unites human beings with a world of purpose. Quine unites human beings with a world of scientific appearances. The world of political purpose and development is consistent with the understanding that human beings speak for their own actions and policy preferences. The world of their experiences and appearances, too, is consistent only with the understanding that they, when they use their language or express their preferences, speak about their developmental aspirations. That most political preferences continue to be justified on the basis of the logic of policy choices seems reason enough for the national interest to continue its entire range of STD public policies and their institutions regardless of the present status of knowledge within a given boundary. If scientific life could take place only in a void, it would have no practical value. If, on the other hand, the only theory of political preference and scientific action rested on the denial of the possibility of determining human purpose, we would have no less a problem. With Aristotle, we would still be political experts; with Quine, scientific heretics; with neither, human beings.

CONCLUSION

This review of the status of science and technology and forms of developmental theory and policy research suggests that a development-oriented government whose programs are carefully coordinated, and whose processes for generating and evaluating research information are efficient and accurate, and whose STD policy goals and objectives are clear and coherent, can achieve an accelerated STD cost-effective infrastructure. Institutional depletion in this context will be low. A development-oriented government whose development policy intents, goals and objectives are deficient in these respects will be inclined, instead, to be erratic in its STD policy formulation and operating responsibilities. Development wastage in this environment will be high.[30]

Therefore, the key to a successful STD development-oriented policy lies in the nurturing and improvement of the government's management of its disparate STD institutions. The provision of the tools that are needed to reduce policy duplication, increase program monitoring and policy development, and promote greater operating efficiency, as well as providing institutional autonomy, would have direct impact on governmental operations and its STD future. The authors of the other pieces in this volume, each in his own way, reiterates and re-emphasizes these points over and over again.

Notes

1. Richard A. Garver (ed.), *Research Priorities for East Africa* (Nairobi, Kenya: East African Institute Press, 1966), see also the publication of the East African Academy, *Research Services in East Africa* (Nairobi, Kenya: East African Institute Press, 1965).
2. Richard J. Barber, *The Politics of Research* (Washington, DC; Public Affairs Press, 1966), pp. 71–90. See also Mekki Mtewa (ed.), *Science, Technology and Development* (Lanham, MD: University Press of America, 1982) pp. 69–83.
3. W. W. Rostow, *The Stages of Economic Growth* (New York: Cambridge University Press, 1960).
4. André G. Frank, *Capitalism and Underdevelopment in Latin America* (New York: Monthly Review Press, 1967).
5. Albert O. Hirschman, *The Strategy of Economic Development* (New Haven, CT: Yale University Press, 1958).
6. Economic Commission for Africa, *Integrated Approach to Rural Development in Africa* (New York: United Nations, 1971); and see *Africa's Strategy for Development in the 1970s* (New York: United Nations, n.d.).
7. Richard Rose, *Managing Presidential Objectives* (New York: The Free Press, 1976), pp. 145–69.
8. Arnold Rivkin, *Nation-Building in Africa: Problems and Prospects* (New Brunswick, NJ: Rutgers University Press, 1969), p. 156.
9. Ibid., p. 157.
10. James Pickett, 'Development planning in Ghana', *Economic Bulletin for Africa*, Vol. 12 (1976), pp. 9–18.
11. F. O. Masakhalia, 'Development planning in Kenya in the post-independence period', *Economic Bulletin for Africa*, Vol. 12 (1976), p. 25.
12. Andrew M. Kamarck, *The Economics of African Development* (New York: Frederick A. Praeger, 1967), pp. 209–21.
13. O. B. Forrest, *Financing Development Plans in West Africa* (Cambridge, MA: MIT Center for International Studies, 1965).
14. Albert O. Hirschman, *The Strategy of Economic Development* (New

Haven, CT: Yale University Press, 1958); also his 'The search for paradigms as a hindrance to understanding', op. cit.

15. W. A. Lewis, 'Aspect of economic development', background paper for *The African Conference on Progress Through Co-operation* (Kampala, Uganda: Makerere University College, 1965). (Mimeographed.)

16. Peter M. Blau, *The Dynamics of Bureaucracy: A Study of Interpersonal Relations in Two Government Agencies* (Chicago: University of Chicago Press, 1969), p. 257.

17. Victor A. Thompson, *Bureaucracy and Innovation* (University, AL: University of Alabama Press, 1969).

18. Emmett S. Redford, *Democracy in the Administrative State* (New York: Oxford University Press, 1969), pp. 52–3.

19. International Bank for Reconstruction and Development, *Seminar on Consulting Services* (Washington, DC: IBRD, 1972), pp. 1–27.

20. Edward J. Schumacher, *Politics, Bureaucracy, and Rural Development in Senegal* (Berkeley, University of California Press, 1975), p. 86.

21. Robert S. Friedman, *Professionalism: Expertise and Policy Making* (New York: General Learning Press, 1971), pp. 1–22.

22. Dale D. McConkey, *MDO for Non-Profit Organizations* (New York: AMACOM, 1975), pp. 99–114.

23. Charles L. Schultze, *The Politics and Economics of Public Spending* (Washington, DC: Theookings Institution, 1968), pp. 51–2.

24. Lyman W. Porter, Edward E. Lawler III and J. Richard Hackman, *Behavior in Organizations* (New York: McGraw-Hill, 1975), pp. 242–5.

25. Mekki Mtewa, *Public Policy and Development Politics: The Politics of Technical Expertise in Africa* (Lanham, MD: 1981), Chapter 1, pp. 1–16.

26. David Easton, *A Systems Analysis of Political Life* (New York, John Wiley, 1965).

27. Mtewa, Public Policy and Development Politics, op. cit., p. 14.

28. Dye, Thomas R., *Understanding Public Policy* (Englewood Cliffs, NJ: Prentice-Hall, 1972), p. 27.

29. Willard van Orman Quine, 'Two Dogmas of Empiricism', in *From a Logical Point of View* (New York: Harper & Row, 1953).

30. Mekki Mtewa (ed.), *Science, Technology and Development: Options and Policies* (Lanham, MD: University Press of America, 1982), pp. 43–80.

Section II
Educational Policy

2 Educational Ideologies and Technical Development in the Third World

Mehrangiz Najafizadeh and Lewis A. Mennerick

Third World countries – whether in Asia, the Middle East, Africa or Latin America – continue to be subject to considerable political, economic and cultural influence by the Western industrialized countries. This influence is manifested in various forms, a paramount one being the spread of Western-oriented education which began during the colonial period and which has intensified since the end of World War II. Although the various cultural and geographical regions of the Third World are diverse in many respects, they are interrelated in that Western-oriented education continues to play a major role in the social and technological transformation of many Third World nations.

More specifically, educational development efforts in many Third World countries have been greatly influenced by what may be referred to as the 'Western educational ideology'. This ideology consists of a set of education-related beliefs and values that define what Western industrialized countries perceive as the 'appropriate' content and scope of education. Central to this ideology is the assumption that Western socio-political values and Western modernization and technological development are appropriate for all Third World countries and that Western-oriented education is a crucial mechanism for achieving such modernization and development. It is suggested, in contrast, that more consideration should be given to the possibility that alternative educational goals and methods may be more appropriate for some Third World countries, and that understanding Third World educational development requires a more thorough recognition of the education-related ideologies that underlie and guide educational development. By focusing on both Western educational ideology and on alternative ideologies, it is stressed that educational expansion should be guided by those particular ideologies that both shape and are shaped by broader socio-political and moral-religious considerations.

SOCIO-CULTURAL VARIATIONS AND EDUCATIONAL NEEDS

Private and governmental agencies in Western industrialized countries have tended to view educational development as a problem that can be solved merely through monetary assistance and through the transfer of conventional Western educational goals and methods to Third World countries. The Western development approach, in its simplest form, has contended that if particular educational models and curricula have been effective in expanding the availability of schooling in the US and Europe, those models and curricula (or variations thereof) should be equally effective in the Third World. For example, if co-educational schooling models are suitable for Western children, they should also be suitable for children in Muslim countries, even though traditional Islamic values prescribe separate schooling for male and female youth.

Thus, Western-oriented educational development programs have reflected (Western) ethnocentric views of education. These programs have not paid adequate attention to the social and cultural variations that exist both between the industrialized West and the Third World, and among Third World countries themselves. They have not paid sufficient attention to the fact that many of the people who are the target of educational development perceive educational needs in ways that differ considerably from those prescribed by the Western education ideology. Western industrialized countries often have attended to the self-interests of the Third World élite while neglecting the desires of the less powerful segments of Third World countries. Although there may be international consensus as to the importance of education in general, considerable differences of opinion continue to exist as to exactly what are the relevant educational needs of the Third World.

Focusing on two geographical areas, Central America and the Middle East, such diversity of opinion is illustrated by the educational change that has occurred in Nicaragua and Iran since their respective political revolutions in 1979.[1] Such change illustrates the necessity of making the content of education more relevant to and consistent with the needs and values of Third World citizens as those citizens themselves perceive their needs. Nicaragua and Iran differ in political and economic systems, and in culture and religion. Yet, prior to their revolutions in 1979, both countries shared particular inter-regional similarities. Both Nicaragua and Iran were governed by dynastic and repressive regimes that were supported by the West generally and by the United States specifically. Further, 'élitist modernization', through which education was primarily available only to urban and upper-class youth and thus was unresponsive to the needs of the general populace, prevailed in both countries. With the revolutions in Nicaragua and Iran came a redefinition and refocusing of educational development reflecting *alternative* educational ideologies. In Nicaragua, this included a mass literacy crusade and efforts to emphasize indigenous values so as to make schooling more nationalistic and more responsive to the actual day-to-day needs of Nicaraguans. In Iran, educational change entailed the purging of Western secularism from education and efforts to develop schooling which reflected Islamic religious beliefs and values.

Both countries, in refocusing education, placed major emphasis on consciousness-raising, on the rejection of dependency on the industrialized West, and on the re-assertion of nationalism and indigenous values. Educational change in Nicaragua and Iran has been guided by alternative educational ideologies that have emphasized the reformulation of the content of education and alternative means of achieving modernization.

This leads to the extremely thorny question: what is 'modernization' and what is 'development'? For many decades, the West has sought to modernize the Third World citizenry according to Western conceptions of modernization. This has entailed the adoption of technology that has originated in the industrialized West and Western forms of education. Concurrently, development efforts also have stressed the adoption of other elements of Western culture. These include Western clothing, art, literature, music and numerous Western consumer items, as well as values and beliefs emphasizing competition and individualism. Educational development geared toward a Western concept of modernization has led many Third World citizens to adopt, to varying degrees, a Western identity at the expense of their own national or cultural one.

Therefore, efforts to develop and modernize the Third World raise a number of fundamental questions. Specifically, what are the desired goals of education, development and modernization? Is technological modernization always desirable? Does technological modernization also require Westernization, or can modernization and development be selective? Can international cultural pluralism be maintained while simultaneously adopting Western technology to improve living standards and social conditions in the Third World? Although it is extremely ethnocentric to believe that the Western life-style is suitable for all world citizens, the drive toward Third World development has been based at least implicitly on just such a view. And Western-oriented education has been central to such development efforts.

The alternative educational ideologies that emerged in Nicaragua and Iran differ from the view that modernization and Westernization are totally synonymous. Educational change in Nicaragua reflects the desire not only to retain Western technology and some aspects of Western culture, but also the desire to emphasize Nicaragua's own unique culture. Educational change in Iran, in turn, reflects the desire to retain Western technology but to reject those aspects of Western culture that are perceived as representing Western secularism.

POLICY IMPLICATIONS

Ultimately, educational development is most fully understood when viewed as part of broader social, economic and political processes. Educational development is neither neutral nor value-free. Instead, it is guided by educational ideologies that both shape and are shaped by broader socio-political and moral-religious ideologies. Educational institutions are not totally independent from other social institutions. And just as the Western industrialized countries use schooling to instill those values, beliefs and skills that they consider important, so do Nicaragua and Iran use schooling to achieve their own forms of socio-political socialization.

Many scholars, policy-makers, and laypersons – both from the West and from the Third World – adhere to the Western educational ideology, and they view it as the only correct approach to Third World educational development. When, as in the cases of Nicaragua and Iran, this ideology is rejected or is not warmly embraced, seldom is there questioning. Instead, the response frequently is simply to conclude that it is these countries and not the Western educational ideology that is in error. Yet, educational change in Nicaragua and Iran suggests that educational policy should be more sensitive to the unique needs of particular countries and more willing to accommodate alternative ways of viewing the scope and form of educational development.

In using educational change in Nicaragua and Iran to illustrate alternative educational ideologies, it is not to be suggested that these two countries are representative of all Third World countries, or that the current Nicaraguan and Iranian educational ideologies are ideal. These ideologies, in the long run, may fail to increase the social well-being of the populace. Certainly, despite substantial accomplishments in reducing illiteracy and in fostering mass schooling, not all Nicaraguans agree that Sandinista governmental policies will ultimately result in improved education and improved social conditions. Likewise, many Iranians who supported the Islamic Revolution had a vision of a 'new' Iran that would have its own national self-identity based on Islamic values and on Iranian cultural values, just as the United States and other Western countries have their unique national identities based in part on Judaeo-Christian values. However, some Iranians now view the Khomeini regime not as an Islamic democracy, but rather as a repressive theocracy, and thus as no better than the repressive Pahlavi monarchy that it replaced. (See, for example: 'Education in Nicaragua,' 1986; Irfani, 1983; 'The New Education,' 1983.)

In this paper the intention is not simply to condemn the Western educational ideology nor to praise alternative ideologies. Western education may continue to be appropriate for the needs of some Third World countries. In other countries, needs may be better met by alternative education ideologies or by ideologies that merely modify Western-oriented education. Nevertheless, attention is drawn to the importance of understanding the ideologies that underlie and guide educational development, and to the importance of recognizing that there is not just one correct set of educational goals and methods. It is imperative to attempt to understand alternative educational ideologies, such as those that have emerged in Nicaragua and Iran, because these ideologies represent a significant form of social change that reflects extremely different conceptions of modernization, of international relations and of the role of education.

Although there are no easy answers to questions about the relationship between education, modernization and development, it is suggested that educational development should rest on certain values that transcend both country and culture. These transcultural values emphasize freedom, self-determination and social well-being of the citizenry. Just as the Western educational ideology should not be imposed on the Third World, through constraint or dependency, neither should alternative ideologies be imposed on the Third World citizenry. In the remainder of this paper four major implications for educational policy are discussed that have emerged from research.

1. How Third World citizens perceive their needs

First, greater attention should be focused on how Third World citizens themselves perceive their educational needs, rather than such educational goals and methods being defined and determined solely by development agencies. Here, the term 'citizens' is used intentionally to emphasize that perceptions of education held by both Third World and Western development agencies sometimes do differ substantially from how Third World citizens themselves define educational needs. Thus, whereas many development agencies traditionally have assumed that there is consensus as to what modernization and technological development should entail and as to how education should facilitate such development, the necessity of recognizing that there are different types and degrees of modernization is stressed, and these may require various forms or models of education. Rather than following a simple diffusionist model of modernization and technological development, greater emphasis should be placed on how particular aspects of Western modernization can be adapted to unique local conditions, and on how the participation of the indigenous population in this process can be increased.

Therefore, greater attention should be paid to the varying needs of *social and cultural regions* where inhabitants share particular values, beliefs, languages, and/or social and economic activities (see Najafizadeh and Mennerick, 1988a). These regions sometimes exist within countries and sometimes cross national boundaries. In addition to the frequently cited distinction between urban and rural regions, there are ethnic regions, religious regions and economic regions.

These regions, which often are interrelated in one way or another, are significant to educational policy because the inhabitants of different types of regions often vary in how they perceive educational needs and ways of meeting those needs. For example, in some cases, *ethnic* values pertaining to family structure and definitions of sex roles may either support or resist formal education. Likewise, *religious* values can be important in determining what is considered to be the appropriate content of education. Values and needs in *economic* regions also vary such that particular types of education, for example, that are considered appropriate for industrially oriented economic regions, may be viewed as quite inappropriate for subsistence or agro-export economic regions.

To focus on social and cultural regions makes explicit the ideological dimension of educational needs. That is, it calls attention to how perceptions of education vary depending on the value and belief systems prominent in a particular type of region. Further, to focus on social and cultural regions also re-directs our attention. The question is *not* why many Third World countries are educationally less developed. Rather, the question is why *some* people in the Third World have access to relatively high levels of schooling whereas education continues to be unavailable or undesirable for *other* Third World citizens. Thus, it is important for educational policy to be more cognizant of social and cultural variations, and to seek to understand more fully how Third World citizens perceive education in relation to their daily activities and their fundamental values, beliefs and traditions, rather than relying solely on policy-makers' preconceived notions of what is appropriate and what is not.

2. Focus on indigenous education and values

Policy should focus more on how indigenous forms of education and indigenous values might be adapted to the twentieth and twenty-first centuries, rather than continuing to assume that only Western-oriented education and Western values are appropriate. By indigenous education is meant the self-determination of educational goals and methods by the Third World itself. Such education may emphasize either the traditional forms of teaching and socialization that were used prior to the adoption of the Western-education model or contemporary indigenous forms that have modified or replaced Western schooling. Indigenous education also refers to the teaching of indigenous values and culture, regardless of whether this is accomplished using indigenous or Western methods. As such, indigenous education refers both to the method and to the content of education. Frequently, policy-makers have neglected the fact that Third World citizens were educating or socializing their youth long before the current expansion of Western-oriented schooling began in the 1950s, and long before the Christian missionaries and Western colonial educators of the nineteenth and early twentieth centuries. Examples include the Islamic university, *Al-Azhar*, in Egypt and the *maktab* or religious primary-level schools in Iran, as well as less formal forms of education and socialization by family and community (see Arasteh, 1969).[2]

In areas where Western-oriented education is not culturally adaptable or responsibe to local needs, development efforts should focus more on how either traditional or contemporary indigenous forms of education might serve to improve the daily lives of the populace. This includes attempting to increase the immediate social well-being, for example in improved health and sanitation education, as well as attempting to increase the longer-run economic well-being of the lower classes. Further, indigenous education, which emphasizes indigenous values and which entails community participation in setting educational goals and methods, also may have a less tangible but equally important outcome by increasing national and cultural self-esteem.

However, in emphasizing indigenous values and education, it is not to be suggested that such education will necessarily satisfy all Third World educational needs. Nor are we glorifying all that is traditional or unique to the Third World. The Khomeini government in Iran, for example, reinstituted a traditional concept of the role of women, including that held by some that women are genetically inferior to men. Such 'traditional' concepts, whether based on religion or on notions of patriarchy, denigrate women and restrict their freedom both in education and more generally. Whether Western-oriented or indigenous, education should occur in the context of the transcultural values noted earlier: freedom, self-determination and social well-being.

3. Countering Western and Third World self-interests

Policy should focus more on how the self-interests that inhibit educational change can be countered and on how the Third World élite can be encouraged to play a more direct and a more supportive role in educational development. Many Third World countries are faced with major problems associated with low gross national products and international debts, and thus economic concerns exert significant constraints on educational development. Yet, Western and Third World self-interests also frequently inhibit major educational change and are an equally important and perhaps an even more enduring problem.

The industrialized West is faced with a dilemma. If Western

industrialized countries support educational development, they are often labeled as imperialistic and self-serving. But if they do not support educational development, they are accused of failing to fulfill international obligations that the wealthier countries should assist the less wealthy. However, the interests of the industrialized West and those of Third World countries are not always compatible. What is most beneficial for the West is not always beneficial for the Third World. Many past educational development efforts have placed too much emphasis on the interests of the Western donor countries, and thus a more even balance must be struck between these often competing sets of self-interests.

Third World educational development should not be viewed solely as a technical, economic or political problem. Rather, it is also a moral problem. The Western industrialized countries bear a moral obligation to assist, but not to dominate. Third World educational development. This obligation includes being more aware of cultural variations and attempting to accommodate them. Third World countries need greater latitude in experimenting with alternative approaches to educational development, including education that emphasizes cultural and national self-identity, even though such education may be offensive, at least temporarily, to some Western industrialized countries. It is ironic that although the French, the British and the Americans, as well as other Westerners, are extraordinarily nationalistic, they tend to view Third World nationalism as a threat and as being automatically and necessarily antagonistic.

Conflicts of interest, however, are not limited to the West. The self-interests of Third World governments and of the Third World élite (including the social and monetary costs that usually accompany educational development) are also sometimes incompatible with those of the populace of the Third World. The governments and the élite of the Third World likewise bear a moral obligation to pursue more actively types of educational development that will enhance the social and economic well-being of the lower classes.

This obligation does not rest solely with Third World governments. Rather, Third World élites should assume greater responsibility in supporting educational change that will benefit the whole populace rather than merely serve to perpetuate their own classes. Consciousness-raising and the mass mobilization of the various strata of citizens were important factors in the recent revolutions and in subsequent educational change in both Nicaragua and Iran. However, such mass mobilization is likely to be most effective only in the short run, as in the 1980 Literacy Crusade in Nicaragua. In the long run – and for the Third World more generally – continuing consciousness-raising and mobilization of the Third World élite is necessary to encourage and sustain educational development. The private sector wealth *and* power that exists in the élite enclaves within Third World countries is considerable. And there is little prospect for substantial and continued educational development unless the Third World élite, despite their vested interests in maintaining the status quo, play a more active role in promoting social and educational change.

4. Increased technical training

Policy should also place greater emphasis on the expansion of various forms of technical education. The development of technologies and control over the distribution of technology-related products continues to be concentrated primarily in Japan, the United States, Canada, Great Britain and Western Europe. Increased attention should focus on technical training for the Third World citizenry that will promote the emergence of indigenous Third World technological research and development programs and thereby increase Third World technological self-sufficiency. Such technical training should be designed so as contribute both to the development of new technologies and to the adaptation of existing technologies to meet particular Third World needs, whether those needs be industrial, agricultural or consumer-oriented.

Further, for many Third World nations during the past decade, varying types and levels of technological development have had differing consequences. Some countries, such as Nicaragua and Iran, have placed major emphasis on the acquisition and/or development of military technologies, thereby reducing the amount of monetary resources available for social services and other development-related programs. Other nations have emphasized industrial and agricultural technologies resulting in an overall improvement of the social welfare of their citizenry. And still other Third World countries have stressed the utilization of industrial and agricultural technologies without fully recognizing that such technologies can also have unintended and detrimental consequences such as environmental (soil, water and air) pollution resulting from their misuse and overuse. Thus, the need for increased technical education applies not only to increasing the number of specialists such as scientists and engineers, but also to increasing technology-literacy among Third World policy-makers and lay-persons more generally so that technologies and technology-related products can be used in ways that will enhance rather than detract from the social and economic well-being of the Third World citizenry.

CONCLUSION

In summary Third World educational policy should attend to at least four major issues. First, greater attention should focus on how Third World citizens themselves perceive their educational needs, rather than educational goals and methods being defined and determined solely by Western and Third World government agencies. Second, policy should focus on how indigenous forms of education and indigenous values in the Third World might be adapted to the twentieth and twenty-first centuries, rather than continuing to assume that only Western-oriented education, Western values and Western conceptions of modernization are appropriate. Third, policy should focus on how the self-interests that inhibit educational change can be countered and on how Third World citizens, including the Third World élite, can play a more direct role in fostering educational development that will benefit all segments of the Third World citizenry. Finally, policy should emphasize increased technical training to permit the Third World citizenry to participate more fully in the development and adaptation of technologies and to use technologies more effectively to improve social and economic conditions throughout the Third World.

Notes

1. For a detailed analysis of educational change in Nicaragua and Iran, using a social constructionist theoretical perspective, see Najafizadeh and Mennerick. 1988b. For other relevant works, see various issues of *Envio*, *Update*, and *International Iran Times*. Also see Arnove, 1986; Beeman, 1986; Black and Bevan, 1980; Borge, 1985; Cardenal and Miller, 1981; Coraggio and Irvin, 1985; 'Establishment', 1987; Harris, 1985; Hiro, 1985; Hussain, 1985; Ismael and Ismael, 1980; Junta for National Reconstruction, 1983; Katouzian, 1981; Keddie, 1981; Kraft, 1983; Krikavova and

Hrebicek, 1981; Macadam, 1984; Miller, 1985; Ministry of Islamic Guidance, 1982; Mohsenpour, 1988; Shorish, 1988; Sobhe, 1982; Thomas, 1985; Thompson, 1977; Tunnermann, 1980; Walker, 1985, 1986; Zabih, 1982.
2. For contrasting views of whether Islamic schools can adopt and serve modernization functions, see Foley, 1977, and Thompson, 1977. For a study focusing specifically on indigenous values and education, see Nxumalo, 1988.

References

ARASTEH, A. R. *Education and Social Awakening in Iran: 1850–1968*. Leiden, Netherlands: E. J. Brill, 1969.
ARNOVE, R. F. *Education and Revolution in Nicaragua*. New York: Praeger, 1986.
BEEMAN, W. O. 'Iran's religious regime: What makes it tick? Will it ever run down?' *The Annals*, Vol. 483 (1986), pp. 73–83.
BLACK, G. and BEVAN, J. *The Loss of Fear: Education in Nicaragua Before and After the Revolution*. London: World University Service, 1980.
BORGE, T. 'The new education in the new Nicaragua'. In Marcus, B. (ed.), *Nicaragua: The Sandinista People's Revolution, Speeches by Sandinista Leaders*. New York: Pathfinder, 1985, pp. 66–90.
CARDENAL, F. and MILLER, V. 'Nicaragua 1980: The battle of the ABCs'. *Harvard Educational Review*, Vol. 51 (1981), pp. 1–26.
CORAGGIO, J. L. and IRVIN, G. 'Revolution and democracy in Nicaragua'. *Latin American Perspectives*, Vol. 12 (1985), pp. 23–37.
'Education in Nicaragua: More students, but what are they learning?' *Update*. Washington DC: Central American Historical Institute, 23 May 1986.
Envio (various issues). Managua: Instituto Historico Centroamericano.
'Establishment of the University of Islamic Sciences in Qom'. Washington DC: *International Iran Times*, Vol. 17 (24 July 1987), p. 2.
FOLEY, D. E. 'Anthropological studies of schooling in developing countries: some recent findings and trends'. *Comparative Education Review*, Vol. 21 (1977), pp. 311–28.
HARRIS, R. 'The revolutionary process in Nicaragua'. *Latin American Perspectives*, Vol. 12 (1985), pp. 3–21.
HIRO, D. *Iran Under the Ayatollahs*. London: Routledge & Kegan Paul, 1985.
HUSSAIN, A. *Islamic Iran: Revolution and Counter-Revolution*. London: Frances Pinter, 1985.
International Iran Times (various issues). Washington DC.
IRFANI, S. *Revolutionary Islam in Iran: Popular Liberation or Religious Dictatorship?* London: Zed Books, 1983.
ISMAEL, J. S. and ISMAEL T. Y. 'Social change in Islamic society: the political thought of Ayatollah Khomeini'. *Social Problems*, Vol. 27 (1980), pp. 601–19.

JUNTA FOR NATIONAL RECONSTRUCTION. 'The philosophy and politics of the government of Nicaragua'. In Rosset, P. and Vandemeer, J. (eds), *The Nicaragua Reader: Documents of a Revolution Under Fire*. New York: Grove Press, 1983, pp. 254–69.

KATOUZIAN, H. *The Political Economy of Modern Iran: Despotism and Pseudo-Modernism, 1926–1979*. New York: New York University Press, 1981.

KEDDIE, N. R. *Roots of Revolution: An Interpretive History of Modern Iran*. New Haven, CT: Yale University Press, 1981.

KRAFT, R. J. Nicaragua: educational opportunity under pre- and post-revolutionary conditions'. In Thomas, R. M. (ed.), *Politics and Education: Cases from Eleven Nations*. New York: Pergamon, 1983, pp. 79–103.

KRIKAVOVA, A. and HREBICEK, L. 'The educational reform in Iran'. *Archiv Orientalni*, Vol. 49 (1981), pp. 221–39.

MACADAM, C. 'Towards democracy: the literacy crusade in Nicaragua'. *International Review of Education*, Vol. 30 (1984), pp. 359–69.

MILLER, V. *Between Struggle and Hope: The Nicaraguan Literary Crusade*. Boulder, CO: Westview, 1985.

MINISTRY OF ISLAMIC GUIDANCE. *The Dawn of the Islamic Revolution*. Tehran: Ministry of Islamic Guidance, Islamic Republic of Iran, 1982.

MOHSENPOUR, B. 'Philosophy of education in postrevolutionary Iran'. *Comparative Education*, Vol. 32 (1988), pp. 58–75.

NAJAFIZADEH, M. and MENNERICK, L. A. 'Worldwide educational expansion from 1950 to 1980: the failure of the expansion of schooling in developing countries'. *The Journal of Developing Areas*, Vol. 22 (1988a), pp. 333–58.

NAJAFIZADEH, M. and MENNERICK, L. A. 'Defining third world education as a social problem: education ideologies and education entrepreneurship in Nicaragua and Iran'. In Miller, G. and Holstein, J. A. (eds), *Perspectives on Social Problems*, Vol. 1. Greenwich, CT: JAI Press, 1988b, pp. 285–315.

NXUMALO, A. M. *The Indigenous Education of the Swazi and Its Implications for Curriculum Development*. Unpublished doctoral dissertation. University of Kansas, Lawrence, 1988.

SHORISH, M. M. 'The Islamic revolution and education in Iran'. *Comparative Education Review*, Vol. 32 (1988), pp. 58–75.

SOBHE, K. 'Education in revolution: is Iran duplicating the Chinese cultural revolution?' *Comparative Education*, Vol. 18 (1982), pp. 271–80.

'The new education in Nicaragua: an open debate'. *Envio*, No. 22. Managua: Instituto Historico Centroamericano, April 1983, pp. 20–7.

THOMAS, R. M. 'Political rationales, human-development theories, and educational practice'. *Comparative Education Review*, Vol. 30 (1985), pp. 312–20.

THOMPSON, A. R. 'How far free? International networks of constraint upon national education policy in the Third World'. *Comparative Education*, Vol. 13 (1977), pp. 155–68.

TUNNERMANN, C. *Hacia una Nueva Educación en Nicaragua*. Managua: Ministerio de Educación, 1980.

Update (various issues). Washington DC: Georgetown University Central American Historical Institute.

WALKER, T. W. (ed.), *Nicaragua: The First Five Years*. New York: Praeger, 1985.

WALKER, T. W. *Nicaragua: The Land of Sandino* (2nd edn). Boulder, CO: Westview, 1986.

ZABIH, S. *Iran Since the Revolution*. London: Croom Helm, 1982.

3 A Research Model Applied to a Computer Project in Swaziland

Martha Tyler John and Floyd Idwal John

INTRODUCTION

In less developed countries there is a need for improved training in science and technology, but resources for providing this training are limited. This lack of resources is particularly evident in fields that are highly dependent on computers and computer-related technology. Few teachers have had experience with computers, most students have never seen one, and computers are not available for use in schools generally. Limited resources in computers, and a lack of teacher training in the field of high technology are generic problems in developing countries. The Kingdom of Swaziland is no exception.

Even though the number of computers available is limited, the computer 'revolution' has begun in Swaziland. A survey conducted by the authors (1984) indicated that there were over 100 computers being used in business, government agencies and schools throughout the country. It was estimated that probably an equal number of computers were owned personally or were already on order and by now this number has no doubt increased significantly.

Computer experts envisage a wide range of benefits for developing countries based on computer use. The training of more science and mathematics teachers, the facilitating of more efficient service in businesses and even the ability to write and use logic are all benefits to be derived from a computerized society. The services provided by computers would be most beneficial to Swaziland where urbanization and population growth are taking place rapidly.

Pupils in the primary and secondary schools had almost no contact with computers, and yet these pupils were the very population from which the country must draw its pool of workers as expanding technology demands new skills. The University of Swaziland had several Apple IIe computers by 1982, and a small number of students

43

were gaining knowledge and skill in operating them. The Ministry of Education and educators generally recognized the need to train a larger population to serve the increasing demand for computer usage. Computer education at the secondary school level would have been ideal, but the cost of equipment for such training was prohibitive.

Then, in 1982, one secondary school, Waterford Kamhlaba, received a grant from an outside agency to stimulate increased use of computers in the school. The objectives of the grant were to:

1. provide a quality of instruction that would allow for long lasting pupil involvement in computer related areas;
2. provide a partner working relationship where pupils work together and reinforce each other as they learn; and
3. 'teach students to think independently'.[1]

The objectives seemed to state desired outcomes, but some system for gathering data to assess the degree to which these objectives were achieved was needed. The remainder of this study examines the effectiveness of the computer training program that was initiated at Waterford School.

BACKGROUND INFORMATION

The computer training program that was developed at Waterford Kamhlaba Secondary School in the Kingdom of Swaziland was a unique and innovative project to be undertaken in a developing country. The Director of the project said, 'We are only too aware that we are setting off into virtually virgin territory,'[2] But the idea behind the project was a powerful one indeed, and so it was undertaken. This underlying idea is stated well by Seymour Papert, one of the project's supporters, who says in his book, *Mindstorms*,[3] '. . . the child programs the computer and, in doing so, both acquires a sense of mastery over a piece of the most modern and powerful technology and establishes an intimate contact with some of the deepest ideas from science, from mathematics, and from the art of intellectual model building.'[4] Exposure to computers and instruction in the use of a computer language would undoubtedly benefit pupils in a number of ways including the building of mathematics, science and computer skills. These skill areas are extremely important in achieving and maintaining a reasonable standard of living for the people within the country. 'Mathematics is much more than the manipulation of numbers. At its best, it involves simple, clear examples of thought so apt to the world we live in that those examples provide guidance for our thinking about problems we meet subsequently.'[5] If the computer project that was being envisaged at Waterford could develop this sense of mastery and the ability to generate clear thought processes for the pupils at the school, it would be well worth the effort to develop the curriculum.

1. Cognitive benefits

(a) Problem-solving

In order for the pupils to acquire knowledge, actual experience with a 'high tech – high touch' learning system was needed. The fourteen Apple IIe computers that had been donated were set up in a room that would allow the pupils to experiment with them and work on problems at their own pace. Thus, the opportunity to 'try out' ideas was providing the basic experience that is needed to generate problem-solving and logical thinking. In formal reasoning and problem-solving the first step is the recognition that a problem exists. 'Once the problem is recognized, the problem solver must represent it in a suitable formalism and then plan a course of action using this representation and a knowledge of the effects of proposed actions.'[6] Working with the computer allowed the pupils to 'analyze problems and test possible solutions and attain success in problem-solving and a personal application of new knowledge.'[7] The problem-solvers (pupils) became knowledgeable about the type of information that the computer needed and could decide when the information at hand was inadequate and needed to be supplemented or changed.

(b) Logical thinking

Seymour Papert[8] and others have postulated that computers can be used as the basis of Piagetian learning or learning-without-teaching. 'Piaget views the child as developing increasingly well-articulated and interrelated representations that are used to interpret the world. Interaction with the world is crucial because when there is sufficient mismatch between the representations and reality, the representations are modified or transformed.'[9] In *Mindstorms*[10] Papert also postulated that 'the computer can concretize (and personalize) the formal'. In other words, the computer can be used as an aid in the development from concrete operational thinking to formal thinking, as an interactive tool that allows knowledge to be explored in a more concrete way. Although some Form I students at Waterford were older than 12 (Piaget's age for formal thinking), it is possible that some of them had not yet reached the formal thinking stage. If there were such pupils, the computer should have been valuable in aiding them to become formal, logical thinkers.

The use of the computer can improve the pupils ability to think logically and to communicate this thinking to others. In comparing the programmer with an expository writer Van Dyke[11] sees several similarities. 'Both must try to anticipate and prevent undesirable consequences – the programmer by providing for all possible logic paths, the essay writer by forestalling, as fully as possible, the reader's misunderstanding.'[12] Students can develop cognitive skills by critiquing and revising the structure of their own and other pupil's programs. The ability to communicate a series of logical thoughts effectively on the computer is no less a task than communication in any other language.

(c) Creativity

The term creativity in its most simple definition deals with the capacity for making something new. It can refer to a person, product or process. Creativity has to do with exploring the unknown, but there are several ways in which one can explore. Almost every human endeavor requires some form of creative output. The areas of ideational fluency (the ability to produce many ideas) and flexibility (the ability to produce different types of output) are inherent in computer programming, particularly in the LOGO programming language. This language lends itself to producing a variety of creative outputs using a parameter varying (i.e. trial and error) approach. Specialists in the area of creative thought indicate that it is possible to stimulate adolescent creativity by using imaginative methods and materials and by actively involving students in trying new ideas. It is also useful to provide a receptive, encouraging atmosphere so that creative ideas can flow. Surely, instruction and practice in the use of the LOGO computer language would qualify as an innovative and imaginative approach to instruction. If this is true, then, the pupils' creative output should increase with the use of the computer.

J. P. Guilford[13] indicates that ideational fluency underlies other aspects of creative behavior; for example, problem-solving ability appears to be related to ideational fluency. There is also a relationship between ideational fluency and the use of heuristic or trial and error approaches to problem-solving. Michael Wallach[14] concluded from his studies that the parts of the Torrence Tests of Creative Thinking that were most likely to evaluate creativity (apart from intelligence) are ideational fluency and fluency-related forms of originality. When pupils are allowed to explore options and design programs themselves they become fluent with the language of the computer itself. Student designed programs may 'provide challenges often screened out of introductory assignments: exception handling, unambiguous and fail-safe user interfaces, and control of logic. . . .'[15] The student learns to manipulate ideas and to apply them in concrete programs that are understandable.

2. Attitudinal benefits

In using the computer pupils who have little or no experience with technology develop attitudes which can play an important role in their later lives. Students' attitudes can have a bearing on their performance in school, and their longer lasting interface with the computer world once they have completed formal schooling. 'Attitudes and values are based on both the information the individual has assimilated and the emotional response specific stimuli evoke.'[16] The pupils who are working '. . . with an electronic sketchpad are learning a language for talking about shapes and fluxes of shapes, about velocities and rates of change, about processes and procedures. They are learning to speak mathematics, and acquiring a new image of themselves as mathematicians.'[17] This learning can create a very positive image of the self and in the process involve the emotions of the student. It seems clear that assimilation of information and emotional response are inexorably intertwined in the process of working effectively with computers and computer programming. Travers says that information can be provided for people in an effort to change attitudes. 'The information may then be forgotten, but it has already had its impact on attitudes.'[18] If students in Swaziland were to learn to use computers and to continue this learning on their own in the world of work, they would most certainly need to have positive attitudes toward this general area of learning.

HYPOTHESES

Several ideas may be derived from the background information presented here. The benefits of computer use having been explored, the following hypotheses were designed to provide feedback about the effects of the project at Waterford Kamhlaba:

1. If pupils from Waterford and two control schools are given logical thinking tasks prior to receiving computer instruction, there will be no significant difference between groups.
2. If pupils at Waterford are given instruction in LOGO language and in computer usage, they will produce significantly more ideas as measured by an Ideational Fluency Test than those pupils who have not had the treatment.
3. If pupils at Waterford are given instruction in LOGO language and in computer usage, they will produce ideas on a wider range of categories are tested by 5 sub-tests of an Ideational Fluency Test than those pupils who have not had the treatment.
4. If pupils are given instruction in using the computer and the opportunity to explore alternatives, they will show significantly more positive attitudes toward computers, mathematics and languages than those pupils who have not had the treatment (as measured by a Semantic Differential instrument).

METHOD

It was important to gather baseline data on the effects of the use of computers on pupils who had had little or no exposure to this type of technology, as well as on pupils who were directly involved in this innovative project. The study was designed to do this as follows.

1. Population

Three schools were selected for the study. The schools are all in urban or suburban areas in Swaziland. One school, Waterford Kamhlaba, is near Mbabane (the capital city); the second school, Nazarene High School, is in Manzini; and the third, Swazi National High School, is in Matsapha (adjacent to the University campus). Although the schools are in urban/suburban areas, many students who are boarders at the school come from rural village areas.

Students at the three schools come from many different types of home. Waterford is a private school which admits students primarily on the basis of academic merit irrespective of national background, color, creed or parents' ability to pay private school fees. About 40 per cent of the students receive scholarships to pay their expenses and in 1983 approximately 35 different nations were represented at the school. The entering pupils are in Form I, and all the instruction is in English. Students take courses in English, French and a third language which depends on the student's mother tongue. They also take mathematics, physical sciences, social sciences, physical education and art. Graduates of the school receive some of the top 'O' level scores in Swaziland.

Manzini Nazarene and Swazi National are public schools which admit students with varying academic abilities. Nazarene High School has a restricted admissions policy because of limited facilities. Preference is given to students who have a Second Class Pass or better. Consequently, Nazarene High School is among the top schools in 'O' level results in the nation. Competition for a place at Manzini Nazarene is keen and there are roughly 20 applicants for every space. The subjects that are taught here are similar to those at Waterford except that there is perhaps less emphasis on foreign languages such as French.

Swazi National High School was established by King Sobuza II in an effort to provide more opportunity for all the pupils in the nation to obtain a secondary education. It serves a larger number of students and has facilities for more classes and a wider range of academic abilities, and provides instruction in subjects similar to those that are mentioned for the other schools.

Form I classes were chosen in all three schools. Generally, students had been assigned to the Form I classes more or less randomly, so the population within each class was heterogeneous. Roughly 50 pupils were used for each school sample.

2. Procedure

Initially a donor contacted the Headmaster of Waterford Kamhlaba, Mr Jennings, with a grant that would provide for 14 Apple IIe computers and some teacher training monies. Mr Jennings then contacted the University of Swaziland for assistance in setting up a model for assessing the effects of computer instruction on pupils. The authors were members of the faculty at the University and offered to develop a research model and present it to the administration and faculty of Waterford for their modification and approval. The other school headmasters and teachers were then approached with the idea of testing Form I pupils within their schools and of serving as control populations.

LOGO had been selected as the programming language. It is the language in which 'Turtle Graphics' are used. 'Turtle Graphics form a good vehicle for learning basic programming techniques – use of variables, recursion, subroutines, etc.'[19] Ten teachers received initial training during the week of the 5th–11th September 1982. In January 1983 four staff members visited New York, New Hampshire and Boston to gain further hands-on experience as well as to observe LOGO being used in a variety of settings.

Teachers and administrators then participated in the planning for data collection to begin in late 1983 or early 1984. They assisted in organizing the pupils for the initial testing in November 1983, and in January 1984 the Form I pupils that were to be part of the study were formally tested.

3. Testing

Three types of measure were developed and given to students at the three schools described previously. At each school, the pupils were tested during class time.

(a) Logical thinking tasks

Five Piaget tasks were given to each of the pupils in Form I. The materials used were common to the culture and therefore to the pupils themselves. The following tasks were administered:

1. Conservation of length – 2 roads and 2 matchbox cars
2. Conservation of weight – 2 balls of clay and a kitchen scale
3. Conservation of mass – 2 balls of clay
4. Classification – 2 doll trousers and 4 doll dresses (superordinate category clothes)
5. Displacement of volume – 2 cylinders, 4 plastic film containers.

(b) Creativity tests

The A form of a Split-half Ideational Fluency Test was administered to pupils in January 1984 and the B form was administered in June

Topic	Time	Instructions
Topics	2 min.	List all the ideas you can about 'Crossing A Stream'.
Theme	2 min.	Write all you can about 'A Locked Door'.
Thing category	1.5 min.	The category is 'Blue'. Go ahead and write all things that are always blue or that are blue more often than any other colour.
Plot titles I	1.5 min.	Write as many titles as you can for the following plot.
Plot titles II	1.5 min.	Write as many appropriate titles as you can for the following plot.

FIGURE 3.1   *Summary of ideational fluency test items*

1984 (Correlation between these two halves had been found to be 0.83 in a previous study done by the author.[20]) This test included the following sub-tests:

1. Topics test
2. Theme test
3. Thing category test
4. Plot titles test I & II.

See Figure 3.1 for additional information.

(c) Attitudes measurement

Semantic differential word comparisons were given to determine the pupils attitudes toward computers, mathematics, English and other languages. Special instruction was given to the pupils so that they would understand the way a word pair response should be made. See Figure 3.2a, b for a sample of the test.

RESULTS

Instruments measuring ideational fluency, attitudes and cognitive abilities were used to obtain information about the populations and to determine the results of one group (Waterford) using computers during the term. Both pre- and post-tests were given to measure

	Strongly agree	Agree	No preference	Agree	Strongly agree	
(1) Computers:						
easy to understand						hard to understand
interesting						boring
time saving						time consuming
useless						useful
bad						good
(2) Mathematics is:						
easy						hard
interesting						boring
useless						useful
good						bad
my favourite subject						my least favorite subject
(3) English is:						
hard						easy
interesting						boring
useful						useless
bad						good
my least favourite subject						my favorite subject
(4) Other languages (such as French) are:						
easy						hard
interesting						boring
useless						useful
bad						good
my favourite subject						my least favorite subject
(5) Computers can be used:						
with difficulty						easily
in many ways						in few ways
by few people						by many people
in many countries						in few countries

FIGURE 3.2a   *Semantic differential*

change in ideational fluency and attitudes. The Piagetian tasks were administered at the beginning of the experiment to determine similarities (or differences) between groups. The results of the testing will be discussed in the same order as the hypotheses were stated.

1. Cognitive (Piaget) tasks

To determine support for the first hypothesis, five Piagetian tasks were presented to the students before the computer courses were

			Pre %	Post %
Computer is:	Easy	Agree	52	71
	Interesting	Agree	84	86
	Useless	Disagree	96	95
	Bad	Disagree	88	86
	Timesaving	Agree	68	57
Mathematics is:	Easy	Agree	56	57
	Interesting	Agree	84	81
	Useless	Disagree	96	95
	Good	Agree	86	95
	Favorite	Agree	60	62
English is:	Hard	Disagree	76	52
	Interesting	Agree	68	67
	Useful	Agree	100	95
	Bad	Disagree	96	86
	Least favorite	Disagree	64	52
French is:	Easy	Agree	32	81
	Interesting	Agree	72	90
	Useless	Disagree	92	90
	Bad	Disagree	80	90
Computers can be used:	With difficulty	Disagree	56	57
	Many people	Agree	80	95
	Few	Disagree	76	81
	Many countries	Agree	76	71

FIGURE 3.2b   *Semantic differential*

started at Waterford. This was done to see if there were any differences between the students at the schools as far as their cognitive abilities were concerned. The five tasks were:

1. Classification
2. Conservation of length
3. Conservation of mass
4. Conservation of weight
5. Displacement of volume.

The results of the testing were analyzed to determine whether there were significant differences between sexes. An analysis of variance for categorical data[2] was used to test for sex differences. There were no significant differences between male and female responses at a given school, so the results at each school were pooled.

The overall percentage of students successfully completing each task is shown in Table 3.1. Two sets of results are shown for the displacement of volume. None of the students successfully completed all four parts of this task. Thus, zero is shown for the results in that column. However, during the task completion, some students became

TABLE 3.1   *Piaget tests*

	C	L	M	W	V
Waterford	78	96	94	78	0(24)
Nazarene	78	80	85	85	0(20)
Swazi National	60	42*	66	75	0(12)

*Differed significantly from the other schools at the .05 level.

aware of the fact that displacement of water depends on the volume of the submerged object and not on its weight. Consequently, they gave a correct response to the crucial, last question, even though they missed an earlier one. The percentage of students in that category is shown in parentheses.

For each of the tasks, an analysis of variance for categorical data was used to determine whether there were differences between schools. This analysis adjusts for the different numbers of students of various ages. The students at the three schools differed only in the conservation of length at the 5 per cent level. However, there were consistent patterns in the result as shown in the table.

In general, the first hypothesis was supported since a significant difference was found for only one Piagetian task.

2. Ideational fluency

Pre- and post-test average scores were determined, by sex, for each of the five parts in the ideational fluency test for each of the three schools. A statistical test of the difference between means showed the male/female averages were not significantly different for any of the categories at any of the schools. However, of the 30 possible comparisons between sexes (5 categories at 3 schools for pre- and post-tests), the female average exceeded the male average 26 times so there was a strong tendency for females to produce more ideas than males.

Since sex differences were not statistically significant, the results at each school were pooled to give overall pre- and post-test scores for the five categories and for the total scores. These scores are given in Table 3.2 for pre- and post-tests for all three schools.

Comparisons between schools were made by comparing the mean change (average post score minus average pre score) for each of the categories and for the total score. Based on a test measuring the

TABLE 3.2   *Ideational fluency, Pre/Post scores*

	Topics	Theme	Category of thing	Plot Title I	Plot Title II	Total
Waterford	5.7/6.5	39.2/40.1	4.6/9.0*	3.2/6.6*	3.5/6.0*	56.2/68.1
Nazarene	3.0/3.7	31.4/31.3	5.9/7.0	3.8/4.2	3.7/4.4	48.0/50.6
Swazi National	3.2/3.5	20.8/14/4	4.7/6.2	2.8/4.7	3.4/3.6	35.0/32.1

differences between means, the improvement (or lack of improvement) at the other schools was less marked than the improvement at the experimental school. This supports hypotheses two and three.

3. Attitudinal scale

The results of the (semantic differential) attitude tests were summarized by sex for the pre- and post-tests at each of the schools. The responses for males and females were compared using a chi-square test at each school to see if there were differences between sexes. No significant differences were found, so the results at each school were pooled before comparing pre and post attitudes at the schools. The strongly agree and agree responses were combined to show overall agreement with a statement. Likewise, the disagree and strongly disagree responses were combined to show overall disagreement with a statement.

A summary of the pre and post percentages of agreement (or disagreement) with statements revealed several differences between schools. However, one of the primary reasons for differences between schools was the number of neutral (no preference) responses. Generally, students appeared to give no preference responses if they did not have any direct experience with computers. As a result, students at Nazarene High School and Swazi National High School gave a high percentage of neutral responses because they did not know anything about computers. Because of the large number of neutral responses, it was decided that comparisons between the three schools should be made using only the non-neutral responses.

Table 3.3 gives a summary of the responses and the results of comparing schools, using a chi-square test. It can be seen that in most cases involving computers, students who had opinions gave similar responses at the three schools. The major differences between

TABLE 3.3 *Non-neutral attitudinal responses, pre/post percentage scores*

	Waterford (K)	Nazarene	Swazi National
Computer			
Easy (A)	76/83	86/76	76/79
Interesting (A)	95/90	73/89	88/89
Useless (D)	100/100	100/96	69/59
Bad (D)	100/100	92/93	58/83
Time save (A)	89/92	85/88	81/72
Maths			
Easy (A)	82/82	81/93	94/84
Interesting (A)	91/94	92/95	90/80
Useless (A)	96/95	78/92	73/72
Good (A)	92/100	100/100	91/97
Favorite (A)	68/93	90/93	89/93
English			
Hard (D)	90/65	79/72	48/46
Interesting (A)	77/82	90/86	87/79
Useful (A)	100/100	84/88	83/89
Bad (D)	100/100	90/93	79/74
Least Favorite (D)	84/73	60/71	66/56
Other Language			
Easy (A)	42/85	85/95	95/88
Interesting (A)	82/95	87/96	60/88
Useless (D)	96/95	85/89	60/80
Bad (D)	100/100	84/89	58/73
Favorite (A)	44/93	82/82	87/86
Computer			
Difficult (D)	93/92	42/39	33/39**
Many Persons (A)	100/95	75/91	92/87
Few (D)	90/81	62/54	29/18**
Many Countries (A)	100/71	81/91	93/90

(A) = Agree
(D) = Disagree
** Significantly different at 5% level (Pre attitudes)

the schools occurred in connection with the responses pertaining to the three subjects: mathematics, English, and other languages. Several interesting observations can be made regarding attitude changes toward these three subjects. It appeared that student reactions to individual teachers may have been primarily responsible for some of the results.

TABLE 3.4   *Attitudes towards computers, consistent responses, per cent*

School	Positive	Neutral	Negative	Mixed
Waterford	60	2	10	28
Nazarene	20	22	22	36
Swazi National	22	43	12	23
Combined	31	22	14	33

Of the nine statements pertaining to computers, the responses at the three schools differed during the pre-test for two statements. In the first, 'Computers can be used with difficulty', students at Nazarene and Swazi National tended to agree with the statement, whereas Waterford students disagreed with it. Similarly, the statement 'Computers can be used by few people' produced different results. Most Waterford students disagreed with this statement, while Swazi National students agreed with it and Nazarene students were more evenly divided.

It was apparent that there were probably anomalies in some of the responses; in other words, a student might have a positive attitude towards computers in answering one statement and a negative attitude in answering another. To explore this possibility, a more detailed examination of results was made for the following three statements:

1. Computers are easy to understand.
2. Computers can be used with difficulty.
3. Computers can be used by few people.

Student responses to these three statements were examined to see if the student was consistent; that is, if the student agreed with the first statement, he would probably disagree with the other two. Conversely, if the student disagreed with the first statement, he would probably agree with the other two. If a student gives a neutral response to any of the statements, the student can still be consistent in answering the remaining ones.

The results of the consistency analysis are shown in Table 3.4. It can be seen that Waterford students are strongly consistent and positive in their attitudes towards computers. In the other schools there is still a high percentage of neutral responses or mixed responses (some positive and some negative).

APPLICATIONS FOR EDUCATIONAL DEVELOPMENT

There are a number of problems involved in setting up a computer program such as the one described in this research. Many less developed countries do not have the funds for the number of computers needed for use in high schools. Teachers may lack the training to deal with computer use and the pupils may have unrealistic expectations of computers. For example, many seem to think that the computer runs on its own with little assistance once you plug it into the electrical circuit.

Some of the problems might be avoided if careful planning is done in advance. A list of 'things to do' is provided below to help those school systems that are considering computer courses.

1. Prepare to purchase enough computers so that two or three pupils can work on one computer together in class. Depending on the size of the class, then, 10 to 15 computers would seem to be required. Funding may be sought from outside sources or the Ministry might put aside money to provide for one beginners classroom in one year. This could be tried on an experimental basis and then expanded gradually.
2. Provide formal training for the teachers who are going to be involved in the computer project. Be sure that this consists of hands-on opportunities, not just talk about the equipment. It might be advisable to have a person from outside the school system work on the first training session until the teachers can help each other.
3. Allow adequate time for the teachers to absorb and experiment with the training ideas. If possible, allow the teachers to work with a few students initially, and then build to a total class. This will help get the bugs out of the instructional system with less embarrassment to the teachers.
4. Provide a curriculum that is spelled out clearly for at least the first few weeks of instruction. This should include pre-tests, actual activities for use in class, homework assignments and evaluation strategies.
5. Help the teachers to provide motivation for pupils by arranging opportunities for them to take the pupils 'behind the scenes' in business where computers are being used extensively. Travel agencies, large department stores and banks are possibilities in most countries, at least in the large cities.

Computers are here to stay and they have made many of the activities in our expanding world much easier. Less developed countries can use computers to provide better records for health services, for use in banking, and for many other services. If computers are used to assist countries in serving the people in them, then computer operators will be needed. The pupils in this research project were a small sample of the first wave of young people who will be able to serve their countries by providing service through a fast, accurate medium, the computer.

Notes

1. Cameron, Ann and Richardson, David. *Junior School Logo Project at Wuterfront Kamhlaba UWSCA – An Introduction*. Distributed Paper, p. 2, 1983.
2. Jennings, Athol R. *Computers at Waterford Kamhlaba*. Distributed Paper, p. 3, July 1982.
3. Papert, Seymour. *Mindstorms: Children, Computers, and Powerful Ideas* (New York: Basic Books, 1982).
4. See note 3, p. 5.
5. Lawler, Robert. *Extending a Powerful Idea*. Massachusetts Institute of Technology: Logo Memo 58, p. 1, July 1980.
6. Fischler, Martin and Firschein, Oscar. *Intelligence: The Eye, the Brain, and the Computer* (Boston: Addison-Wesley, 1987), p. 87.
7. See note 1.
8. See note 3.
9. See note 6, p. 133.
10. See note 8.
11. Van Dyke, Carolynn. 'Taking "computer literacy" literally'. *Communications of the ACM* (May 1987).
12. See note 11, p. 371.
13. Guilford, J. P. *The Nature of Human Intelligence* (New York: McGraw-Hill, 1967).
14. Weir, Sylvia, 'Logo and the exceptional child'. *Microcomputing* (September 1981).
15. See note 11, p. 372.
16. Chase, W. Linwood and John, Martha Tyler. *A Guide for the Elementary Social Studies Teacher* (Boston: Allyn and Bacon, 1978), p. 233.
17. See note 3, p. 13.
18. Travers, Robert M. W. *Essentials of Learning* (New York: Macmillan, 1977), p. 506.
19. See note 1.
20. John, Martha Tyler. *The Relationship of Modeling and Imitation to Ideational Fluency in Homogeneous and Heterogeneous Groups*. Doctoral dissertation at Stanford University. June 1966.

21. Light, R. J. and Margolin, B. H. *An Analysis of Variance for Categorical Data in Three Dimensions*. Paper presented at the meeting of the ASA, Montreal, Canada, 1972.

4 Cultural Constraints in the Transference of Computer Technologies to Third World Countries

Elia Chepaitis

In the past fifteen years, computerization has revolutionized the production of goods and services in the North. Newly industrialized and developing countries such as Pacific Rim nations, Brazil and India have moved swiftly and with some success to computerization. Yet computers represent an opportunity *and* a threat to those countries loosely designated as the 'Third World', particularly to the cluster of lesser developed countries (LDCs) which have borrowed heavily.[1]

The costs of acquiring computers are not only economic, but also political and socio-cultural. Campaigns to diversify economies and to achieve domestic reform through computer technologies create fears of a syndrome of spiraling debt and dependence, and also of the destruction of valued socio-cultural mores and traditions without the evolution of acceptable alternatives. The Third World fears Northern dominance through control of four vital resources: finance, management, technical know-how, and information ownership. The revolution in global telecommunications and information services has left Third World countries more dependent upon international data management and with less control over foreign access to domestic information. LDCs especially worry that imported technology will contribute to a decline in national sovereignty, living standards and internal stability.

Costs and risks are substantial not only because sacrifice is necessary to match limited resources with enormous needs, but also because of the swift tempo of change and its instant communication. The hasty purchase of volatile, ill-tested, expensive machines can lead to over-investment, imbalance and waste in any culture.

Inter-regional cooperation and international agencies such as the

61

Industrial and Technological Bank (INTIB) and regional associations disseminate technical information, technical aid and data on negotiations with Northern vendors. Agencies such as INTIB often serve as models for national fact-gathering and advisory agencies. To maximize internal control, Brazil and India have developed their own computer industries. Technology transfers are affected not only by national posture and policy, and the culture of the adoptive nation, but by the interaction of several cultures.

TECHNOLOGY AND CULTURE

Technology and culture are intimately related; computers can threaten culture, and cultures raise physical and attitudinal barriers to the successful implementation of proven technologies. Four cultures affect the transfer of invention and applications:

1. the culture of national origin (such as the United States);
2. the culture of the vendor-producer (such as IBM);
3. the culture of national destination (such as Nigeria);
4. the culture of the adoptive enterprise (such as a government agency or a private business).

Each culture affects and is affected by technology, struggling to optimize invention yet sacrificing efficiency for effectiveness by humanizing the machine to suit local needs and preferences.

Technology has never been culturally neutral. The West compromised rationality in scientific practice, machines and processes to accommodate its heterogeneous cultures, priorities of élites and resource endowments. For example, the United States invested in the steamship and the railroad when the conquest of nature and the frontier was a dominant ideal. In contrast, the bicycle, the automobile and the airplane were produced and marketed generations later, almost simultaneously, when individual mobility and sport became valued.[2] Often, design and application errors are perpetuated by cultural mindsets and the inability or unwillingness to break from tradition; in time, invention can become embedded in culture, although the ethic and myth of inventiveness may be given lip service. American railroad technology, for example, remains inefficient and inadequate; obsolete design and management continues to be supported. Railroads modified nineteenth century cultures, but culture (in part) constrains the perfection of the railroad.

Although the United States created the computer industry, mainstream and organizational cultures engage in an unending cost–benefit analysis of information technologies which include considerations such as planned obsolescence, employee health, access control, depersonalization, and the significance of a growing information society. Some of the stickiest problems and issues in computer systems include:

1. insufficient user-friendliness;
2. unresolved issues of information ownership versus sharing;
3. telecommunications policy;
4. the spiraling costs of maintenance;
5. inadequate training personnel and facilities;
6. conflicts between organizational discipline and creativity in changing job descriptions;
7. inadequate and inappropriate reward systems for users who 'convert' to the new technologies;
8. hardware and software incompatibility, often within the same product line;
9. unreliable connectivity;
10. ineffective security systems;
11. built-in obsolescence and waste.

Excessive costs and machine-related dehumanization have also been related to two cultural biases: a preoccupation with technological virtuosity (i.e. invention for its own sake), and the powerful influence of professional or political interest groups; these groups are, in essence, mini-cultures, formed by cliques of public officials, specialists, the military or intellectuals.[3]

Not only the national culture in which technology originates, but also the culture of the vendor-producer affects design and utilization; producers and vendors often impede the implementation of ideal technologies for non-technical considerations. Hardware and software vendors balance profitability against long-term viability, commitments to customer support, and the need for industry standards. Vendors tailor licensing agreements, user-friendliness and product improvements to match their perceived market niche, mission and resources. For example, Lotus Corporation's customer and production policies vary markedly from Hewlett-Packard's.

Historically, cultures of destination have controlled and enhanced foreign inventions successfully, producing combinations which often seem incongruous to the cultures of origin. The Western bicycle was adapted creatively to create efficient, urban public transportation by the addition of the cart, to produce the labor-intensive but simple, cheap rickshaw.[4] Transistor radios, TVs, automobiles and bootlegged urban electricity are utilized in far flung underdeveloped areas, examples of powerful global pressures to secure and modify Western invention, with or without planning.[5] This technical borrowing often creates technological pluralism between geographic and social units, and also within households. Machines change the culture of the household, and the culture of the household seeks to overpower and control technology.

In adopting computers, the Third World should consider in a cost-benefit analysis the opportunity costs of computerization: would the capital, talent and time that is invested in the transfer be used to better advantage elsewhere in society? On the other hand, what are the opportunity losses if computerization is not achieved in cost-effective and strategic areas? Furthermore, an analysis of the ripple effects of the investment is vital. Structural reorganization, improvements in transport and utilities, and, perhaps above all, enhancements in education and social mobility which accompany the implementation of new technologies may be of greater benefit and significance than computerization itself. The planned adoption and adaptation of appropriate technologies may serve as a powerful vehicle for long-desired social and economic change.

The fourth culture, that of the adoptive enterprise, is shaped by its internal resources and value systems, and also by external interactions with the three other cultures which impact on computer system transfers: those of the technology's national origin, the national destination and the producer-vendor. For example, enterprises such as nationalized units, joint ventures or subsidiaries of an MNE (multinational enterprise) often lack the independence to select balanced, appropriate, competitively priced and expandable technologies primarily for the health of the enterprise. Internally, an adoptive enterprise such as a bureaucracy has a singular life and characteristics. It may be an ancient patrimony, a jerry-rigged service organization staffed by a unique subclass or even foreign nationals, or a unit dominated by military, university or provincial representatives.

PROBLEMS IN THE THIRD WORLD

It is not possible or desirable to separate cultural from political and economic factors in assessing the seminal problems that the Third World encounters in computerization. Power groups, economics, law and external considerations are shaped by culture and each can affect policies and postures which supersede cultural considerations. Each nation's institutions, legal characteristics, resource manipulation and priorities are imperfect and in flux, but they are the unique and logical product of a singular heritage, environment and pattern of interactions. An appreciation of the *logic* of culture within a current political and economic context is invaluable in assessing problems and possible solutions in specific technology transfers.

The appropriate cultural components to examine in a discussion of Third World cultures are:

1. attitudes and beliefs, including religion;
2. social institutions;
3. the material culture, including the infrastructure;
4. education;
5. language;
6. aesthetics.

Several attitudes are commonly singled out in international business courses as potentially incongruent with machine technologies. Significant cultural dislocation often is attributed to mechanization, when in actuality the attendant urbanization and changes in lifestyle are also culpable. Typical areas of conflict with traditional views and mores are:

1. industrial discipline;
2. punctuality and scheduling;
3. planned systems of waste;
4. monotonous tasks;
5. the devaluation of craftsmanship;
6. remote decision-making;
7. indifference to seasonal imperatives and traditions;
8. expectations that management has external patriarchal duties;
9. the lack of prestige for production workers.[6]

Two additional attitudes specifically affect computerization:

1. a romanticization of computers;
2. an excessive fear of computerization.

When technology is valued for its own sake and when computers are mistakenly regarded as omnipotent 'thinking' machines, cultural dissonance, unrealistic expectations, excessive costs and a lack of control follow. Although technological naïveté is often associated with the working class, it is those in decision-making positions – political, military and business leaders – who shop mythologically.[7] They commonly seek status, the 'quick technical fix', and permanent solutions to badly analyzed problems through costly and uninformed buying sprees from a catalogue.

On the other hand, a culture rejects computerization as an insidious vehicle of despised Westernization at their own risk. The rise in religious fundamentalism, and its disgust with intrusive foreign goods and values, is a logical outgrowth of the swift globalization of economies and consumer aspirations. It is difficult to see how cultural integrity can be preserved without technical-economic leverage and competitive machine solutions to dynamic global change. Historically, the Middle East and China were most culturally secure when they were technologically superior to the West.

Social institutions such as the family, village or free associations may impair technology transfers because of their influence on worker mobility, educational opportunities and group values. Labor unions and consumer organizations are often powerful forces which may react negatively to 'foreign technologies', perceiving threats to nationalism, employment and a familiar lifestyle. Group attitudes toward working women, women's education and women operators of powerful machines may also contribute to a reluctance to accept foreign technologies lest they 'Westernize' traditional family structures and increase demands for foreign goods and services.

Material culture affects the potential for successful technological transfers in visible and quantifiable ways. Transfers are facilitated by efficient transport, mature communications and stable electrical utilities. The condition of the material infrastructure may determine when, where and how computers are installed, and may require substantial investments and design changes before computerization can take place. Uncontrollable physical factors such as climate, topography and seasonal changes also require hardware adaptations, to deal with humidity, dust or extreme temperatures. Transport and energy, as well as comparative advantages in factor endowments influence the choice of production and information systems.

Experiences with unsuitable capital-intensive systems have led to unemployment, debt and instability and have created awareness of the important questions which must be asked of vendor and producers. The Third World is increasingly aware of the desirability of simple machines, easy repairs, affordable financing, appropriate and main-tainable software, and access to source codes and updates. Although the effort of melding material culture and imported computers can be strenuous, hard data is increasingly available for informed decision-making and planning in this area. Case histories of systems selections in developed countries, particularly by small businesses and public entities, are frequently surprisingly relevant for LDCs. In addition, domestic economic statistics, such as GNP/capital, population growth and the distribution of wealth help to indicate which level of technology is affordable.

Mass literacy, mass education and ongoing training are prerequisites for the effective and timely utilization of computer systems which will yield a rich social dividend. From robotics to data entry, more and more education is required to unleash the power of the best technologies. A file cannot be accessed, an operation cannot be specified, manuals are in effect unavailable, without a literate oper-ator. Adult literacy facilitates balanced and broad conversions from manual operations, minimizes generational attitudinal differences. Mass education is vital, to avoid a two-tiered society composed of a mobile technically adept class and a teeming, impoverished subclass. Educational constraints in technology transfers include:

1. imbalanced intraclass educational opportunities;
2. the neglect of apprenticeship and informal training;
3. the reluctance of the vendor to provide sufficient training and training materials;
4. an overconcentration on higher education;
5. the lack of opportunity and incentives to stem the emigration of qualified professionals and managers;
6. a reliance on visiting or resident foreign technicians.

Cultural shifts in both the North and the Third World are necessary to avoid an entrenchment of experts with narrow accountability, loyalties and training. Since the computer industry has matured and products are mass-marketed, the dominance of exclusive technical élites in computer applications is no longer necessary and has proved to be counterproductive, leading to inappropriate, intimidating and disruptive systems.

Language not only is the means of communication, including culture-specified idioms, but also is related to ease of use and true ownership of technology. Manuals can be translated, software can be amended with appropriate commands and menus, and alphanumeric

characters can be altered. On the other hand, every effort should be
made to select publicly tested, off-the-shelf mainstream software for
which customization, documentation, training and upgrades will be
available on a long-term basis at favorable terms. The decision
whether to import foreign technical terms or to insist upon 'native'
words and eschew foreign technical jargon deserves significant attention and analysis.

Language is seminal in culture, but the costs of translating for a
broad range of changing technologies may be steep in the short term;
in the long term, investments in a generation of students trained in
documentation and programming may not only ease the price of
adequate translation, but also lead to the production of appropriate
domestic additions to transferred products. However, when value is
added to the software or to information services by the receiving
cultures, potential profit from reselling the amended product or
services may be sacrificed if the improvements are unavailable in a
mainstream language such as English.

Aesthetics are also significant in cultural expression and communication; cultural considerations such as sound, workspace configuration, color and output formats are desirable for effective systems
designs. For example, users' aesthetic sensibilities may be offended
and productivity may be reduced by auditory prompts, 'busy' screens,
printer clatter, peripheral clutter or resource sharing. The rate of
technology transfer will probably not be affected severely by aesthetics
unless applications involve the design of culture-specific commodities
or the production of computer-assisted art or music.

Finally, attitudes toward information and information handling are
strongly affected by each of the four cultures, including the culture
of the enterprise. Resistance to record keeping, data gathering,
information sharing and global data communication may be well-
founded to protect proprietary or private information. However, if
in the existing manual system information is hoarded for personal
turf-building, financial gain, lassitude or generalized suspicions, often
to the detriment of an organization as a whole, information will
probably not be well-managed and shared in automated systems
without passive or active resistance.

OPTIONS AND SOLUTIONS

The optimal tactic in accommodating cultural constraints is to *adapt*, *adopt* and *adapt*: to adapt computers to local cultures and needs, to adopt the altered technologies, and then to continue to control and enhance the hybrid solutions. However, all parties seldom have the willingness, the resources, the expertise, or the communication to choose this option. Costs, time constraints, technological complexities, and the desire to utilize the automated system as soon as possible create pressures which often encourage short-term solutions which clash with culture and common sense; in addition, the system is unlikely to be implemented efficiently or effectively. Some nations, such as the People's Republic of China, purposefully bypass the possibility of local adaptation, purchasing plug-in turnkey systems intact from the industrialized North.

In general, however, the Third World will hybridize suitable technologies. The Third World can avoid repeating the mistakes of the past, including past patterns of technological misdirection – pursuing invention, development and investment regardless of society's needs. Even if it avoids the mistakes of technical leaders, of course, the Third World is free to make its own mistakes.

Guidelines for appropriate technological transfers which minimize cultural dissonance should include:

1. short-term (two years) and long-term (five to seven years) strategic planning to coordinate technology with culture, and also to coordinate acquisitions of multiple desirable technologies;
2. a thorough assessment of the costs, opportunity costs and opportunity losses involved;
3. generous and appropriate realistic reward systems for users;
4. an identification of the structural changes required by computerization which are beneficial in themselves;
5. a prejudice in favor of simple solutions complimentary to the factor endowments of the culture;
6. a study of similar implemented systems in similar cultures;
7. the ongoing participation of political and business leaders, specialists *and* users in control and review;
8. broad-based, open and dynamic planning for change and expansion, including domestic enhancements and invention;
9. international cooperation for equitable and honest transfers of equipment and technical knowledge.[8]

In the past five years, more perfect information and more perfect competition improved the output of technical information and of publicized bidding and purchase agreements by organizations such as INTIB. In addition, the increasing availability of expertise in intercultural transfers and the mellowing of extreme attitudes toward computers increase the odds that educated choices may produce optimal solutions. In addition, market-driven improvements in hardware and software in 1990 offer superior performance and ease-of-use at 1980 prices. The advantages of being a technical 'follower' rather than an early leader are significant. The movement toward more reliable hardware, voice systems, software 'shells', and packages which can be easily customized for specific environments benefits the Third World.

CONCLUSION

The first step toward successful transfers of computer technologies is the identification of cultural constraints. The second is to determine which internal changes are desirable, which are unavoidable, and which should be prevented. The third step is to question whether the technology in question must and can be adapted, and whether the hybrid product is worth the long-term cost.

Cultures are inherently conservative. In view of the waste and misinformation involved in computer transfers, this conservatism may not be misplaced and can be channelled. Change is a modern constant, and planned change is a modern imperative for the Third World. There is seldom justification for rushing to acquire state-of-the-art technologies, although it may be desirable to be a well-informed 'early follower' in competitive areas and in proven technologies which improve the general welfare, such as health or distribution systems. An educated and self-conscious effort to minimize intercultural friction and to implement desired technologies is a creative endeavor, releasing a culture's adaptive as well as adoptive capabilities.

Notes

1. Freidan, Jeffrey A. and Lake, David A. (eds). *International Political Economy: Perspectives on Global Power and Wealth* (New York: St. Martin's Press, 1987), p. 201.
2. Pacey, Arthur. *The Culture of Technology* (Cambridge, MA: MIT Press, 1984), pp. 86–7.
3. Ibid., pp. 10 and 81.

4. Rybcyzynski, Witold. *Taming the Tiger: The Struggle to Control Technology* (New York: Penguin Books, 1985), pp. 207–8.
5. Ibid., pp. 195–227.
6. Ball, Donald A. and McColluch, Wendell H. *International Business* (Plano, TX: Business Publications, 3rd edn, 1988), p. 193.
7. Heinrich Klaus, 'Technological Assessment: An Essentially Political Process', in *Impact of Science in Society*, Vol. 36 (1), 1986, pp. 65–76.
8. For an exploratory reading, consult El Sawy, Omar A., 'Implementation by Cultural Infusion: An Approach for Managing the Implementation of Information Technology', *MIS Quarterly*, 9(2), 1985, pp. 131–40, and Poznawski, Kazimierz Z., 'Technology Transfer: West–South Perspective', *World Politics*, 37(1), 1984, pp. 134–52.

Section III
Institutional Policy

5 Experts, Advisers and Consultants in Science, Technology and Development Policy

Mekki Mtewa

'Give us a guide,' cry men to the Philosopher. 'We would escape from these miseries in which we are entangled. A better State is ever present to our imaginations, and we yearn after it; but all our efforts to realize it are fruitless. We are weary of perpetual failures; tell us by what rule we may attain our desire'.

'Have a little patience,' returns the moralist, 'and I will give you my opinion as to the mode of securing this greatest happiness to the greatest number.'[1]

INTRODUCTION

Instead of helping to solve the development puzzle, experts, advisers and consultants tend to confuse and compound it. Unless a developing government has a prior understanding of the role of these advisers, it cannot use this very expensive manpower resource effectively. This concern, therefore, raises two issues in development policy. The first relates to a government's manpower policy and how it addresses short-term, highly professionalized consultants. To benefit from this temporary help, should not a developing government include among its permanent staff members possessing the equivalent or corresponding professional skills? In making this determination, it must be presumed that a developing government knows what professional skills to acquire and how to match its advisers with maximum efficiency. It is also imperative that a government knows what policy structures to establish within which these interactions could take place.

The second concern relates to the clarity of development policy itself and its relation to the professional manpower component. Here, I am concerned that advisers and consultants should be involved throughout the political organization in order to have a comprehensive impact on the development activities of the government. Failure to provide a structure within which permanent government workers and paid consultants can work together produces disinclination toward such help in governments which could most benefit from it.

In assessing the development planning experiences of Cameroon, Malawi, Tanzania and Zambia, it may be concluded that the failure of all the First Five-Year Development Plans rendered these governments averse to the aid of experts.[2] However, the critique fails to depict the state of development research, the planning techniques in use and the database in existence in these developing countries. Lack of proper research facilities or accurate data is disadvantageous and discouraging to those experts who cannot work effectively without them.

Contrary to these admonitions and negative expressions directed toward experts in Africa, their use has increased. Developing governments regard the use of these professionals as part of the solution to manpower shortage, and part of the national, technological by-product of organizational efficiency.[3]

Gerald Meier predicted that the volume and quality of project experts who come to Africa vary with their duties. Much of their consultancy and professional work involves warning the new states 'what will happen in the economy if no changes are made, and reveals what can be made to happen by policy measures that control certain strategic variables.'[4]

Morris Davis reveals that in the western region of Nigeria, the British firm of Patrick Dolan and Associates served as consultants to the western regional government of Chief Obafemi Awolowo. Besides providing such public relations staples as advertisements, brochures, pavilions and press conferences, their work included efforts to improve or design indigenous strategies, advice on legislative tactics, and the lobbying of members of the British Parliament. During the 1959 election they were active in tactical and strategic campaign planning.[5]

The evidence provided by Davis reveals some hidden forces behind the nationalist capabilities in developing countries. The advisers responsible for coordinating the lobbying efforts of the nationalist movements probably know more about their developing clients than the leaders themselves. In the United States, the Foreign Agents

Registration Act of 1938, Section 11, requires these agents to register with the Office of the Attorney General.[6]

The Malawi government has also used foreign experts, advisers and consultants to help determine its development ambitions. In January 1965, Messrs Brian Colquhoun and Partners, a British firm, were appointed to conduct feasibility studies of the New Lilongwe Capital City project.[7] No European government was, at that time, interested in funding this project. In October of the same year, a public administration specialist working with the Economic Commission for Africa assessed the administrative and efficiency requirements of the Malawi government in relation to the existing Zomba capital.[8] By December 1965, the South African government was identified as a possible funding source. However, South Africa needed substantiation of the feasibility studies by its own national experts. With the help of Dr P. S. Rautenbach, the South African Prime Minister's Planning Adviser, the Johannesburg consulting firm of Messrs W. J. C. Gerke and Withers were commissioned to conduct this study.[9]

With the master plan brought in to supplement the reports and surveys already available, the South African Minister of Finance announced on 27 March 1968 the granting to Malawi of an 8 million South African rand loan at 4 per cent interest. To ensure the security of this loan and maximum use of South African materials, the loan stipulated that the New Lilongwe Capital City Project be contingent on the maximum use of South African contractors and of South African materials in the case of imported requirements.[10]

This illustration raises the question of developing countries' use of different experts to satisfy different parties whose technical assistance objectives are contrary to each other. Experts who have no distinct governmental ties are concerned only with a client's ability to pay the agreed fee and on time. If a client can do this, the bond of expertise between the two parties is cemented. Experts working for the bilateral and multilateral donors, however, are more concerned with meeting the requirements of their principal donor. The interests of the recipient government are secondary to the fulfilment of the terms of his contract.

DEFINITION

The word 'expert' is a noun derived from the Latin *expertus*, meaning one who has tried, proved and known by *experientia*. *Experientia*, the Latin root of our English 'experience', establishes a clear linkage between the profession of the expert and the asset or major tool of his trade. Experience is such an asset to the expert as prudence is to the statesman. Experience, in the original Latin version, implied a form of balance in professional growth – a balance which ensured a communion between the mere substance of knowledge, its application, and the time over which one has consistently been accurate.[11]

An 'adviser', in the Latin root, is one who advocates or proposes laws, gives counsel, or interprets things of divine will. According to Machiavelli, an adviser is necessary because too often statesmen do not understand what it is that they see. Statesmen may be endowed with conceptual thought, but the relation of this thought to effective truth could be defective. The adviser, presumably, possesses an analytic mind capable of analyzing and ordering intuitive evidence into coherent sets of historical information.[12]

'Consulting', on the other hand, is an organized effort by specially trained and experienced persons who have demonstrated some depth in knowledge and scope of solving problems. In Latin, the consultant and the adviser are one.[13]

Contemporary experts, advisers and consultants have a certain presumptuousness that derives from a conviction that modern administrative and intellectual techniques permit them to act in a variety of fields in which they not only have limited knowledge and experience but are even too ignorant or unpracticed to be sensitive to the gaps in their capabilities.[14]

Hendrik S. Houthakker writes of the advisory problems of the Council of Economic Advisers in the United States: 'As a member of the council, I was involved in problems ranging from copper pricing to farm policy, from supersonic transport to textiles, and in many others. In retrospect it seems clear that I was doing too many different things, far more than I could have any adequate knowledge of. My only excuse for staying with most of them anyway was that the other people involved often appeared to know even less than I did, or else had an axe to grind.'[15]

Guy Benveniste rebuked those who profess expertise but are unfamiliar with the fundamentals of the profession itself. 'Men little think how immorally they act in rashly meddling with what they do not understand,' Edmund Burke said. 'The expert is attacked,' Benveniste repeated after Burke, 'because she/he appears unable to deal with many essential qualitative issues. The ideologies rediscover that rationalization is by definition a schema or simplification, and therefore fails to deal with emerging values.'[16]

Clearly – and quite contrary to current requirements – experience implies those conditions which might be beneficial to experts, advisers and consultants. In so far as experience implies command of knowledge, it is imperative that it intimate some deference from intuitive statesmen. They will continue to generate beneficience from their clients comparable to that of Machiavelli's *The Prince*. The primary significance of this assertion is the view that, deficient or not, they perform useful services in technological societies. They do so by providing a rational organizational synthesis to disjointed developing systems. It is not as important to them that they evaluate their own images against their usefulness in areas in which they are the acknowledged authorities. The substance of authority and knowledge becomes, and is, the two elements that contribute to the confidence of their profession.

COMPARABILITY OF OBJECTIVES

The objectives which experts, advisers and consultants set for themselves are, of course, different from those of their clients. The intersection of their objectives is incidental to the tasks, requirements or constraints put on their professions in the attainment of their client's objectives. Six objectives are discussed below which they service within technical assistance. Each objective has a philosophy peculiar to its achievement.

1. The humanitarian objective

This objective is concerned with the needs of mankind and the alleviation of poverty, hunger and disease. The United Nations Charter refers to humanitarian rights in Articles 1(3), 55(C), 62(2), 68 and 76(C). The setting of standards for the achievement of the humanitarian objective is left to individual governments, however. In 1948 and 1966, the Universal Declaration of Human Rights was adopted and amended to include an International Convention on Economic, Social and Cultural Rights.[17]

Critiques of the humanitarian objective imply that its open-endedness leaves governments with discretionary views on its implementation. Developing governments have neither a clear conception of the humanitarian objective nor a definitive answer or methodology for the resolution of its fundamental concerns.[18]

2. The political objective

This objective pertains to the related or unrelated power motives of two political parties, one stronger, the other weaker. The stronger contributes to the strengthening of the political structures of the weaker, thereby justifying why the objective is in their mutual interests. More often, political relationships exist between unequally endowed parties, either in land mass, or population size, or the total sum of their GNP, or defense capabilities. Implicitly or explicitly, political objectives are processes by which states are functionally or ideologically separated.

For instance, in the International Institute of Agriculture, voting was determined by membership in one of five classes, members of Class I having five votes and members of Class V having one vote. Equality was admitted in the sense that each State was free to choose the class to which it wished to belong, but membership in Class I involved an assessment of sixteen units of the budgetary base and membership in Class V involved the assessment of only one unit.[19]

Proceeding from a colonial point of view, a colonial power and its dependency cannot be, but are in law, juridically equal; they can neither enjoy the same political rights nor have equal capacity in their exercise. To what extent do experts, advisers and consultants serve this invidious political objective, and how many of its sentiments do they themselves share?[20]

3. The economic objective

This objective develops out of the need to recognize the fact that the mutual interests of two or more parties lie in the strengthening of their productive capabilities according to the economics of scale.[21] 'States come into existence because no individual is self-sufficing,' Socrates said. So, having all these needs, we call in one another's help to satisfy our various requirements. . . . The conclusion is that more things will be produced and the work be more easily and better done, when every man (or State) is set free from all other occupations to do, at the right time, the one thing for which (it) is naturally fitted.[22]

The imperative of specification as formulated in Plato's *Republic* is now strengthened by theories of value. Adam Smith, in *The Wealth of Nations*,[23] attempted to solve the paradox of value in the economic and trade relationships of states. In Chapter 5 Smith asserted that value is the real worth of a thing to the person who has acquired it, and who wants to dispose of it or exchange it for something else. . . . The power which that possession immediately and directly conveys to him, is the power of purchasing; a certain command over all the labor, or over all the produce of labor which is then in the market. His fortune is greater or less, precisely in proportion to the extent of this power.[24]

Article 1(1) of the General Agreement on Tariffs and Trade, adopted at Geneva on 30 October 1947, acknowledges the paradoxical relationships that have historically existed between and among states and recommends that any contracting party to any product originating in, or destined for, any other country shall be accorded immediately and unconditionally to the like product originating in, or destined for, the territories of all contracting parties.[25]

When the 'Most-Favored-Nation' clause was formulated, 'discrimination' in commercial matters was the general rule among nations. The paradox of value between commodities and the appropriate relationship of labor to its product is compounded by the problem that the most-favored-nation clause has made the establishment of equitable trade agreements between the capital and the labor-abundant economies difficult.[26] What, then, is the economic objective between states?[27] Perhaps the success of the North–South dialogue will provide a key to this question.

4. The historical objective

This objective is the most invoked, but a readily misconstrued, variable in the relationship between states. The idea of 'history', according to Machiavelli, ought to be the basis for political obligation. 'Whenever I have been able to honor my country, even when burdensome and dangerous to me,' Machiavelli contends, 'I have done so willingly, because Man has no greater obligation in life than that: we being dependent for our existence on it, and then all that good which fortune and nature have granted us; and they come to be greater in those men, who have the more noble [kind] of country.'[28]

History is a process by which states establish bonds of friendship and dependence. Conventions are by-products of familiarity with each other's expectations and needs. Conventions supplement the unwritten laws that regulate these established bonds and expectations. Historical conventions, according to Hugo Grotius, are standards by which states adopt a formula for the restitution of damages caused to, or injuries suffered by, one another. Appropriately, historical conventions are 'promises made, through signature to treaties or otherwise', to which the party states obligate themselves.[29] By contemporary standards and needs, how appropriate and equitable are invidious historical practices that are couched in historical and conventional terms? What revisions should be made to this principle without severing and violating standards of conventional decency?

5. Legal objectives

These objectives refer only to prior, or existing, obligations such as are recognizable or enforceable either by law or force. Legal objectives constitute the first admissible evidence for determining the structure of inequality between states. The essence of the legal objective, Britain once argued, is that all states enjoy an equality of rights but this does not mean that they have the same rights. Every state to some extent circumscribes, or increases, its rights and duties by the treaty commitments into which it has entered.[30]

Bilateral legal objectives achieve and promote interests central to the two states. Whatever the dimensions of their identifiable objectives – technical assistance, trade, culture, or friendship – the legal document sets the terms of reference to each of these arrangements and the means for resolving disputes. Article 14 of the Covenant of the League of Nations provides that if any dispute whatever should arise between (states) relating to the interpretation or the application of the provisions of the Mandate, such dispute, if it cannot be settled by negotiation, shall be submitted to the Permanent Court of International Justice.[31]

In *Duff Development Co. Ltd.* v. *Government of Kelantan* (1924), AC 797, the British House of Lords, in the words of Viscount Finlay, commented that it is quite consistent with sovereignty that the sovereign may in certain respects be dependent upon another Power; the control, for instance, of foreign affairs may be completely in the hands of a protecting Power, and there may be agreements or treaties which limit the powers of the sovereign even in internal affairs, without entailing a loss of the position of a sovereign Power.[32]

Is the legal objective responsible for the vicious circle of inequity? Using equity, can political relationships be more equitable between cooperating states?

6. The egalitarian objective

This objective advocates the doctrine of equal political, economic and legal access to ways and means of self, or collective, authentication. Egalitarianism is not only a historical process but also a social one since its uniqueness clearly lies in the way property and labor promote or stifle its progression. Egalitarianism and a country's mode of production are related since the mode of production of material life determines the character of a country's social, political and spiritual processes of life. It is not the consciousness of men that determines their existence, but on the contrary, their social existence that determines their consciousness. From forms of development of the forces of production these relations turn into their fetters. Then comes the period of social revolution. With the change of the economic foundation in a social revolution the superstructure is more or less rapidly transformed.[33]

Like the Marxist conception of proper relationships of forces of production, egalitarianism aspires to achieve individualized work, price, salary and living conditions for all elements involved in the production process. The International Covenant on Civil and Political Rights of 16 December 1966 and Articles 1 and 55 of the United Nations Charter apply this egalitarian principle to states. States, according to Article 1(2) thereof, may freely choose their modes of production 'without prejudice to any obligations arising out of international economic cooperation, based upon the principle of mutual benefit, and international law'.[34] Between these two dimensions of the egalitarian objective, Marx is the undisputable advocate for individual egalitarianism, and the United Nations for that of states. These two views appear irreconcilable.

TECHNICAL ASSISTANCE OBJECTIVES

In 1965, President Kwame Nkrumah affirmed the view that 'foreign capital is the instrument of exploitation and not development' whereas 'aid is merely a [form of] revolving credit.'[35] These assertions can be evaluated within the context of our six objectives.

Technical assistance is both an old and a relatively new phenomenon. Through the Marshall Plan, the United States of America made a commitment to European governments to aid them in the reconstruction and rehabilitation of their economies. If treated within the comparability of objectives, the interests of the United States and those of the European countries were similar. The aftermath of the Second World War saw the European economies terribly disrupted and in need of reconstruction and rehabilitation funds. The favorable US balance of payments situation played a major role in the United States' decision since it involved replenishing the huge balance of payments surplus it had accumulated.[36]

Colonialism created for Britain and some Western countries, with reference to the Berlin Conference of 1884–5, responsibilities for the peoples colonized.[37] Independence for Europe's colonies, compounded by a lack of capital and technology, did not relieve the former colonial powers of their prior legal, political, economic and historical commitments. Technical assistance becomes an effort on their part to contribute to the restitution of the devastated colonial economies.[38] However, unlike the Marshall Plan, the colonial heritage is an extended and, therefore, most illusory form of technical assistance.

Developing governments compete against each other for technical assistance funds. The use of experts, advisers and consultants within their policy-making machineries of government can increase a government's technical sophistication with the effect that its funding applications may reflect an acceptable style and format and its logic may be better formulated. Eugene Black strongly urged that, in order for developing governments to succeed in obtaining project funds from the World Bank, they should use advanced project analysis methodologies. Failing this, the Bank feels obliged to perform these cursory tasks for them, at a fee.[39] In recent years, the technical veto power of the IMF and the World Bank has grown proportionate to their emphasis and use of technical data unavailable to their beneficiary member states.

Developing governments now use experts, advisers and consultants to help them compile technical assistance reports and funding documents which require statistical and technical sophistication. This increased use of such professionals is directly related to their determination to save time and minimize expenditures on proposal writing. Therefore, the experts, advisers and consultants who serve the policy interests of developing governments create within the structures of these governments a sense of policy integration and community. These integrative functions operate on the horizontal, vertical, and value levels.

The horizontal function is the degree to which members of a national policy-making system have facilities for communications and transactions with other members who hold corresponding roles in the stratification system.[40] Considering in this light the policy-making and communication orientations of development experts, advisers and consultants, one might assume that the extent of interaction between these individuals in the effective pursuit of their duties correlates positively with their mutual inputs into their clients' respective policies. These inputs, in turn, determine the final content and strength of the technical assistance request packets.[41]

A vertical function is the degree to which policy actors in different strata are interlinked.[42] From the perspective of professional utilization, vertical integration narrows procedural, technical and administrative gaps between these organizational experts, advisers and consultants. The process of vertical integration is enhanced by more effective communication. Increased communication, on the other hand, reduces the processing time which a technical assistance agency normally takes to reach a funding decision. The underlying cause of frustration in technical assistance project financing, for example, is the time it normally takes between submission of funding requests and the final receipt of approval on them. Warren F. Ilchman and Norman Uphoff correctly argued that these frustrations have negative investment and performance effects on the developing administrative infrastructures since the time and costs incurred in project proposal writing are 'a fundamental investment aimed at reducing the cost of putting policies into effect.'[43]

The value function here means the degree of value congruence in, and within, the professions that are responsible for the translation, weighting and selling of technical assistance preferences of their developing clients to a wider number of agencies. Values among experts, advisers and consultants emerge as a result of their professional orientations or ethical standards. Members of these professions, although they must at times serve diametrically opposed clients, subscribe to similar functional and structural norms which legitimate their professional authority.[44]

Contrary to Morrison and Stevenson's evidence, value integration in the development planning and implementation process is enhanced by the extent to which professionals of diverse persuasions perceive and respect the professional products of each other within this matrix of professional norms.[45]

A conflict of values between policy-making structures within a development government may pose a number of problems. The first problem may relate to the dispersion of values. The most obvious source of this value conflict is multiplicity of technical assistances and diversity of values in them. The political and professional interests of their personnel component may intersect. However, the central authority may not define the proper relationship of each of these interests to its national objectives.

Several problems arise here. Tamar Golan's study of L'Ecole Nationale de Droit et d'Administration (ENDA) in Kinshasa addresses the first one. A Zaïrean educational project was funded by both the Zaïre and the French governments; this financing would have given it a bilateral character. To complicate the issue, the Ford Foundation, a private United States corporation, was the third financier. The French thereby lost their favorable bilateral treatment. The problem then emerged: because the Director-General of ENDA was so preoccupied, the Frenchman and American who shared power with him were afforded much autonomy. The conflict between them was pervasive; no member of the school community was ever unaware of it. The American was Secretary-General of the school and had been supplied to ENDA by the Ford Foundation. The Frenchman, on the other hand, was a *fonctionnaire* of the French government. Their differences did not paralyze the school, but they did badly complicate its operation.[46]

The second problem concerns the meeting of minds. Uma Lele discusses this in the context of the World Bank experiences. The World Bank administers projects in developing countries where indigenous skilled manpower does not exist. It is the Bank's policy in these cases that, if the process of growth is to be stimulated and the rural development projects are to be sustained, governments must assume administrative responsibility. To illustrate this dilemma, Uma Lele refers to the Lilongwe Land Development Project in Malawi where: 'There was no obvious choice as to which levels should be strengthened to take over responsibility for coordinating its activities: the regional and district level administrations have traditionally coordinated the work at the local level of a number of technical and social service ministries, and may by virtue of this be most suitably equipped for ensuring coordination in Lilongwe in the future. By contrast, however, since services to agriculture have been the central focus of LLDP activities, it may instead be more appropriate to strengthen the agricultural divisions to perform the main coordinating role. More problems arise in the transfer of responsibility for specific elements of the project, especially where more than one ministry should logically be involved.'[47]

The cases of Zaïre and Malawi demonstrate the need for advisers and consultants to be dispersed throughout policy-making structures rather than placed merely at their apexes. This might be accomplished by using the vertical and horizontal dimensions as a guide to allocate experts, advisers and consultants on a project-by-project basis. Such a proposal conforms to what Margaret Wolfson calls 'counterpart staffing', a process by which, based on her experience in Tunisia, 'the artificial insemination project specify(s) that there should be Tunisian counterparts for the Belgian Director and his assistance'.[48] Counterpart staffing demands symmetrical or equivalent operational skills of the project professionals.[49]

Another solution has been to centralize all professionals in their present positions under the Ministries of Development Planning. According to this view, both the Ministry of Development Planning and the experts, advisers and consultants in their respective policy orientations are controlled by the central authority of the Office of the President. Although popular by the standards of British project administration, this approach is deficient according to the Americans. The British and French use it because it serves four of the objectives stated above (the historical, the political, the economic and the legal) to the detriment of the humanitarian and egalitarian aspirations of these old dependencies.[50]

By the implicit invocation of doctrines of colonial privilege and economic interdependence between Britain and France, on the one hand, and the respective dependencies on the other, tension builds between technical assistance programs in Africa and limits the freedom with which governments can freely choose. Fear of colonial foreclosure which Britain and France possess by virtue of their substantial investments there further complicates the contradictions of colonial aid and the contingents of experts, advisers and consultants who manage these interests.[51]

British experts, advisers and consultants come under three technical pseudonyms. The first is the British Overseas Service Aid, established in 1961. The second is the British Aided Conditions of Service Scheme, developed to strengthen the terms and conditions of service prescribed in the first scheme. The third, the British Expatriate Supplementation Scheme concerns the numerous 'supplementation' allowances that accrue to recruits who agree to serve British interests in its independent Commonwealth. The net effect of these three schemes is essentially to 'top up' the local salaries to bring them roughly in line with British salary levels, to provide educational and passage allowances, gratuities (in the case of contract officers), and pensions and compensation payments (in the case of permanent and pensionable officers). In addition, supplemented personnel have longer leave entitlement and enjoy larger housing subsidies than their local counterparts (at the cost of the recipient government).[52]

The three schemes were, in fact, originally introduced into Africa to supplement the Three-Fifths Rule. In 1946, Britain, faced with a war economy and a manpower shortage in the aftermath of World War II, commissioned a study of the colonial service conditions in its African dependencies. In 1946, the Commission that preceded the Lidbury Report argued that there should be a means of inducing British personnel to serve British interests in its Empire. To this effect, the Commission recommended that, since the war had diminished the technical manpower at the disposal of government service and as competition for the existing skills was greater, the British Colonial Office should adopt an economic formula in the form of the Three-Fifths Rule. Essentially, this deals with two separate sets of market values. The prospective candidate from the United Kingdom or elsewhere in the Commonwealth outside East Africa looks first at the salary which he would expect to earn for comparable employment in his own country. He then considers the special conditions of overseas employment, such as the breaking of ties with his home country, climatic risk, loss of amenities and the additional cost of taking a family, and includes, in these days of rapid political development, an element of career insecurity.[53]

The argument is sound that overseas employment should reflect comparable and competitive levels and standards of remuneration. Methods by which governments or private industry can compete for scarce technical resources are matters that can be determined by the market itself. The rules of the market can change but not the preferences of those who possess these skills. British authorities understood how to achieve a competitive salary both domestically and abroad. Abroad, the three-fifths rule established a successful differential among its British nationals, the indigenous personnel, and the nationals of other competitive governments which served in the different programs there. Presently, with its Empire declining, Britain has sought to maintain an acceptable level of confidence among its nationals by requiring that a recipient government employ British officers (from its own funds) to supervise the welfare and general conditions of service of its expatriate nationals and unconditionally accept the schemes in their entirety.[54]

How successful is the United States technical assistance program and its experts, advisers and consultants under these circumstances? Some 29 United States government agencies and bureaus are involved in bilateral assistance operations in Africa. United States bilateral expenditures to Africa represented about 4.61 per cent of the 1976 budget of $6,355,233,000. In 1977, this increased to 5.22 per cent of the estimated $8,176,396,000 budget. The 1978 assistance budget, however, reflected a decrease to nearly 5.3 per cent of the estimated $10,265,077,000.[55] In the fiscal year 1979, the Carter administration asked Congress for almost $294 million in bilateral development assistance for Africa. The future of United States and Africa relations, according to the US Senate Subcommittee on African Affairs, was said to be dependent on this volume.[56]

The documents defining US technical assistance are numerous. However, the terms of reference were first enunciated by President Truman on 12 March 1947, when he declared: 'It must be the policy of the United States to support free peoples who are resisting attempted subjugation by outside pressure. . . . Our help should be primarily through economic and financial aid which is essential to economic stability and orderly policy processes.'[57] And on 20 January 1949, President Truman revealed the source of this confidence: 'For the first time in history, humanity possess the knowledge and skill to relieve the suffering of [developing] people.'[58]

Most recent clarifications of the nature, scope and content of this technological confidence appear in the International Development Act of 1961. The act stipulates that US technology is for the exploitation of resources that have a direct impact on human needs. The Foreign Assistance Act of 1967 amended and significantly expanded the language of this act to include provisions of technical experts to aid developing governments in effectively coordinating their assisted programs. Regarding the issue of jurisdiction, the legislation expressly states that the development enterprise in Africa should be local and not be dictated by any external power. This reaffirmation of the Cold War philosophy has earned the United States a unique reputation as a competitive source of technical assistance that is free from colonial stipulations.[59]

The new International Development Cooperation Act of 1978 expands the Act of 1967 and further incorporates a number of evaluative criteria. Under this Act, the President is authorized to

request, accept or reject technical assistance funding if the performance record(s) of the recipient government is contrary to the humanitarian expectations of the United States and those which appear in
the United Nations Charter. Section 211 of the Act prescribes for
the President eight evaluative questions:

1. whether the activity gives reasonable promise of contributing to
   the development of educational or other institutions and programs
   directed toward social progress;
2. the consistency of the activity with, and its relationship to, other
   development activities being undertaken, or planned, and its
   contribution to realistic long-range development objectives;
3. the economic and technical soundness of the activity to be financed;
4. the extent to which the recipient country shows a responsiveness
   to the vital economic, political and social concerns of its people,
   and demonstrates a clear determination to take effective self-help
   measures and a willingness to pay a fair share of the cost of
   programs under this title;
5. the possible adverse effects upon the United States economy,
   with special reference to areas of substantial labor surplus, of the
   assistance involved;
6. the desirability of safeguarding the international balance of payments position of the United States;
7. the degree to which the recipient country is making progress
   toward respect for the rule of law, freedom of expression and of
   the press, and recognition of the importance of individual freedom,
   initiative and private enterprise; and
8. whether or not the activity to be financed will contribute to the
   achievement of self-sustaining growth.[60]

The United States AID executes its technical assistance program
through a network of country missions, experts, advisers and consultants, and other contracting arrangements. Terms of service and levels
of remuneration are unique to the service component that each
professional group provides. However, nowhere in the documents is
there mention of compensation based on potential loss of leisurely
amenities (such as golf, tennis or the opera) or against such oddities
as climate or the remote possibility of encountering a tsetse fly.
Experts, advisers and consultants who agree to serve the American
interests in Africa perhaps go there with full knowledge of the
perils that exist on the continent. However, unlike the British, the
Americans adopt a realistic view of the African world that there

can neither be an adequate measure of insurable interest nor the extraordinary use of technological equipment that would assure these individuals a stable state of mind. Risk is, therefore, an assumed element in any human endeavor which brings people into contact with each other and sends them beyond their own boundaries.

Governments, then, are free to structure their technical assistance programs in a manner that best translates their objectives. Technical assistance programs of any two governments may or may not use experts, advisers and consultants who share common assistance values. In the case of the British and French, they perceive assistance as a source of colonial influence. The Americans, on the other hand, look at assistance in the humanitarian and/or egalitarian senses in that they postulate assistance as a means by which developing governments should construct a viable take-off economy. These bilateral perspectives, incidentally, operate as follows, alongside multilateral programs that embrace broader values.

1. The World Bank and its affiliates

The World Bank is a specialized agency of the United Nations. Its Articles of Agreement came into force on 27 December 1945, following the Bretton Woods Conference. The International Finance Corporation was established by articles of agreement which became effective on 20 July 1956. The conceptual objective of the International Finance Corporation (IFC) was to further economic development but, unlike the World Bank, to encourage the growth of private enterprise in developing economies. This corporation is an affiliate of the World Bank although it has a separate charter and its own operational funds (Article 4, Section 6). The International Development Association (IDA) was, similarly, established by articles of agreement drawn on 1 February 1960, and which became effective on 24 September 1960. Like the International Finance Corporation, the International Development Association operates as a separate entity from the World Bank (Article 6) and is an affiliate of the Bank. The objectives of the International Development Association are to promote economic development, to increase production and to raise standards of living by providing financing on terms more flexible than those of conventional banks.

The Bank's philosophy regarding Africa is that the leadership in African development belongs to Africans themselves; a development program is meaningless unless it is inspired, framed and supported by the people it is designed to benefit. The Bank's role is essentially no more than that of catalyst. It is each country's sovereign right to decide whether or not a particular policy should be adopted, a particular measure introduced, or a particular project launched.[61]

To increase the leadership potential of developing countries, the bank provides consultative services through which developing countries conduct project feasibilities, appraisals, engineering, resource surveys and training. In fiscal year 1977, 'technical assistance components were included in 151 lending operations, for a total of $230 million, compared with 162 operations, for a total of $189 million' in 1976.[62] Typically, the bulk of the Bank's consultative services call for small teams of planning advisers, supporting consultant services, and training programs for local counterpart staff. Unlike other executive agencies, however the Bank handles the procedural, administrative and other aspects of its role as executing agency on a project-by-project basis.[63]

In fiscal year 1977–78, the World Bank financed approximately 138 development projects of diverse quality and character. The total investments of all these projects totaled $6,097,700,000.[64] By the middle of the 1978–79 fiscal year, the Bank had already agreed to fund approximately 708 development projects entailing funds far in excess of the 1977–78 figure.[65] A total of 25 African countries received 215 (nearly 30.1 per cent) of the projects; another 190 projects (27 per cent) were distributed among 15 Southeast Asian countries. Of the remaining 303 projects, about 164 of these went to 18 European, Middle Eastern and North African countries; and the last 139 projects went to 23 Latin American and Caribbean countries.[66]

The Bank's Project Preparation Office at its headquarters advances monies to developing countries conducting feasibility studies. The borrower repays these advances by refinancing them through another Bank loan or an IDA credit for the project concerned as soon as project funds becoms available.[67]

Experts, advisers and consultants under the World Bank program are identified, screened and assigned on a competitive basis. First, the Bank has criteria by which it measures professionals and ranks them accordingly to how suitable they seem within the Bank's technical needs. A consultant's registry is kept at the bank to which constant reference is made by potential users – both government and statutory bodies.[68]

The Bank does not use experts, advisers and consultants to solve administrative or coordinative problems of its financed projects.

Rather, it views such professionals as a preliminary source of expertise which could prescribe a diagnosis in the form of a feasibility study. On the basis of this feasibility document, the Bank enters into financing negotiations with the potential recipient knowing that the document is not a conclusive basis for finalization of the funding agreement.

The relationships that the Bank's experts, advisers and consultants establish in their preliminary assignment(s) with potential recipients do not impose upon them any additional responsibilities other than to listen to answers which they themselves formulate in the process of their research. As a research effort, it supplements the non-existent or inadequate research manpower at the disposal of the potential recipient. The terms of service and levels of their remuneration reflect the importance the Bank attaches to their research help. Since the Bank makes its final financial decisions on the accuracy of these preliminary projections, it stresses the importance of demonstrative research experience in its research experts, advisers and consultants.

2. The United Nations Development Program

The apparent perfectionism of the Bank is equally true of the United Nations Development Program (UNDP). The Program, established in 1966, has about 244 professional personnel in Africa of whom about 31 are Africans. Local staff in Africa total about 1,029.[69] In 1977, the United Nations Development Program employed 308 consultants. During the first ten months of 1978, there were 500 consultants, most of whom were nationals from developed countries.[70]

The presence of consultants, advisers and experts within the operations of UNDP was originally intended to supplement the technical skills that were lacking within its professional personnel. The role that experts, advisers and consultants may play in UNDP activities has been a central issue of debate within the organization. Recently, the UNDP preoccupied itself with questions of the appropriateness of such services. In addition to insisting on higher professional standards, UNDP is required to ensure that 'the provision of expertise [be] more responsive to the needs of developing countries, as well as more economical.'[71]

In the case of the UNDP, the issue of the relevancy and appropriateness of advisory, expert and consultative services raises questions of interpretation of its original authority. The original intention of the use and role of experts in the UNDP was to be related to the technical and professional requirements of its development assistance programs. Like the World Bank, however, the UNDP programs were to be managed by a resident team. The UNDP's concern in this context is the integration of experts within the resident teams. In addition to this concern, UNDP's problem relates more to making its financed experts, advisers and consultants responsive to the needs of the developing governments themselves than to serving the values of the resident teams. It is a question both of methodologies and policy orientation of the resident teams and the recipient governments. Resolution of these concerns does not rest with the UNDP alone, since it cannot mandate what methodologies experts can and should use. The issue of professional quality and performance is not at question here. One must ask, however, whether governments which have become accustomed to one set of experts, advisers and consultants (such as the British or the French) can have confidence in American or Soviet professionals whom the UNDP uses without discrimination.

3. The United Nations Technical Assistance Program and OPEX

The United Nations, on the other hand, also operates an Operational and Executive Personnel to Overseas Territories program (OPEX or OPAS). The OPEX program was initiated as a result of extension of the activities of the UN Technical Assistance component by General Assembly Resolutions 1256 (XIII) and 1946 (XVIII). The program was intended to provide the emerging states with minimum executive and administrative personnel. Additionally, the United Nations intended for these two resolutions to meet some of the related expenditures of the acquisition of international experts.[72]

In 1975 and 1976, there were only 38 OPEX professionals. In 1977, this figure dropped to only 29, with 30 in 1978. Presently, OPEX operates in Africa, Latin America, the South Pacific, and some Asian states. Africa has consistently received a substantial number of OPEX experts – more than Latin American and the South Pacific combined.[73] Within South Africa, Lesotho, Swaziland and Zambia, there have been higher OPEX allocations than in Botswana, Mali, Sierra Leone and Ethiopia.[74]

The United Nations' OPEX program provides for the appointment of about 100 experts to posts in the civil services of developing governments. Specifically, their assignments are not of an advisory character. OPEX experts receive the equivalent of local salaries of comparable rank. The United Nations pays stipends and allowances as appropriate to supplement their local salaries.[75]

The qualification standards of OPEX experts are high. Those experts who make the OPEX ranks are generally men and women who possess not only high professional competence but also a strong sense of dedication to the aims of the United Nations. Previous experience in developing countries is, of course, an advantage but not a prerequisite to appointments unless specified. United Nations experts must be able to establish friendly relations with the milieux in which they live and work. They should always remember that they are first and foremost members of the staff of the United Nations dedicated to its purposes and ideals: betterment of human conditions without regard to considerations of nationality, race, religion or politics.[76]

The majority of the United Nations experts, the program emphasizes, 'are senior people in their professions with sound academic background and solid practical experience. Seldom does a United Nations expert have fewer than ten years' experience in this field.'[77]

The examples of the World Bank, the UNDP and the United Nations Technical Assistance Program reveal that there is a distinct concern with the roles of experts, advisers and consultants in helping multilateral programs reflect a favorable and efficient view of UN operations. The preoccupations of these programs are, therefore, more with professional standards than with elementary techniques of analysis. This emphasis on the highly professional and technical orientation of UN activities diminishes the enthusiasm and interest of its potential beneficiaries. The sophistication of the work pushes the developing governments either into diverting their development funds from agricultural projects to technical and scientific training courses (an unpalatable choice to them), or into increasing their reliance upon foreign private experts, advisers and consultants to translate these statistical projections.

4. The Commonwealth Fund

To contrast the multilateral effort, one might compare the work of the Commonwealth Fund, a multilateral program formed by Commonwealth members with a distinct view of colonial interest, with the programs of the World Bank, the UNDP and the UN Technical Assistance which have broader conceptions of the world.

The Commonwealth Fund is an integral part of the operations of the Commonwealth. It was established in 1971 with the objective that the Commonwealth provide various forms of technical assistance aimed at meeting the developmental needs of Commonwealth member countries. The governments of Britain, Canada, Australia and New Zealand are the main contributors to this fund.

In 1974, Britain accepted in principle the target, advocated in the UN, of increasing net official development assistance to 0.7 per cent of GNP; Canada also accepted the 0.7 per cent target. Australia was committed to reaching this target by the end of the 1970s. New Zealand surpassed this in the 1975–76 financial year.[78]

The criteria by which the Fund measures the usefulness of its technical assistance are: first, that the activities of the Commonwealth reflect a common historical legacy – that of colonialism; second, that technical assistance serve as a mode of capital circulation within this class of states. The argument of transfer of technology is not as important as these two concerns. The Commonwealth Secretariat identifies whose administrative or procedural components require multilateral coordination. The Fund forges a unity between the bilateral and multilateral assistance programs of its developed states and pools these joint resources to achieve the primary colonial objective. For this reason, no distinction really exists between Britain's bilateral or multilateral assistance efforts and the philosophical underpinnings of the Fund's assistance. As the senior founding member, Britain plays the leading role in the fund.

Since the interests of Britain are to provide Commonwealth governments with enough political manpower, it emphasizes that the Commonwealth Fund should identify European resources that have this capability. According to this view, technical assistance includes the provision of people to run the political agencies in areas where there are insufficiently trained local personnel but support for the training of local people is reluctant. To this end, Britain has contributed over 12,000 advisers, experts and consultants and 2,705 volunteers.[79] Total disbursements for technical assistance since 1972 have consistently been above the 60 per cent level. The numbers of professionals to train Commonwealth personnel, however, declined in 1973 to only 14 per cent from a 17 per cent level in 1972. Total disbursements for these training activities has never risen above 18 per cent of its technical assistance allocations.[80]

The activities of the Fund can only expand at the expense of individual bilateral arrangements. However, since bilateral arrangements are strengthened at the expense of the Fund's program, the 1973 report suggests that technical assistance disbursements to all Commonwealth recipients decreased. This happened after a marked decline in the number of the Fund's advisers and educational experts sent abroad, from 1,599 in 1972 to 1,369 in 1973.[83] Canada increased its bilateral allocations for training with the result that, during the same time span, bilateral Commonwealth trainees increased from 2,203 to 2,245.[82]

The Fund's philosophy regarding technical assistance has, therefore, been evolving. However, the political sensitivity of its operations puts it in a potentially explosive position *vis-à-vis* its major contributors. Since 1971, the technical assistance operations of the Fund have been restricted to recruitment of experts of an advisory or operational nature from the major Commonwealth contributors. Recently, recruitment activities have been undertaken within the developing Commonwealth members themselves, in nations such as India and Nigeria which have attained some political prominence within the Commonwealth itself.[83]

The Commonwealth Fund, like the British OSAS or BESS programs, provides experts and educational advisers on a project-by-project basis. The service contracts are contingent on the requirement that the government making the request contribute to the Fund the equivalent of the local salary for the position being filled. The Fund meets all salaries, allowances, travel and other official items of expenditure relating to these assignments.[84] The total sum of a government's contributions to the fund, however, reflects neither the total number of the fund's experts and advisers in its service, nor their composition, since these factors are hidden behind the mask of colonial privilege. The cost(s) of these technical services are compounded within the individual bilateral agreements under different sub-headings which the recipient governments obligingly sign.[85]

The identification of advisers, Commonwealth experts and consultants is done by individual governments through their Overseas Consultants Corps. These corps maintain registries which they share with the Technical Assistance Group at the Commonwealth Fund. When the contributors to the fund are in need of particular skills or want to match those of their nationals with another Commonwealth member, they refer to this registry. If the project merits a collaborative effort, the Fund assumes responsibility under provisions of advisory assistance. The Fund enters into a contract with these advisers and remunerates them at the established Fund's fee structure. There is nothing, however, to stop any of these governments from 'marking up' the Fund's fees to a level which the individual national state member believes its national advisers are worth. Neither the calculations of the Fund nor the total cost of the project include this incidental cost. There is no consultative fee structure on which outside sources can rely.

CRITIQUE

Despite the minute volume of foreign experts, advisers and consultants assigned to bilateral and/or multilateral programs abroad, these professionals wield immense veto power based on their technical knowledge. This power is hidden behind the force of the programs themselves since it is the professionals who evaluate potential success or failure of their programs based on evidence from the field. Their perceptions are a significant contributing factor in the bilateral and/or multilateral funding of a development program.

Developing governments may have a lot to do with the perceptions of experts, advisers and consultants, culminating in the quality and/or quantity of positive and/or negative recommendations these professionals dispatch to their respective bilateral and/or multilateral agencies. The code of conduct for the US Technical Assistance Program provides guidelines for experts, advisers and consultants on what not to do. However, it is not an absolute standard of impartiality since perceptions are functions of political, historical or other ideological factors. Article 111(2) of the UN Technical Assistance Program reads: 'The Officer shall conduct himself at all times with the fullest regard for the aims of the Organization and in a manner befitting his status under this contract. He shall not engage in any activity that is incompatible with the purposes of the Organization or the proper discharge of his duties with the Government. He shall avoid any action and in particular any kind of public pronouncement which may adversely reflect on his status, or on the integrity, independence and impartiality which are required by that status. While he is not expected to give up his national sentiments or his political and religious convictions, he shall at all times bear in mind the reserve and tact incumbent upon him by reason of his status'.[86]

Developing governments accept multilateral aid and its accompanying experts, advisers and consultants with the belief and conviction that they are useful to their development efforts. The United

Nations Technical Assistance Program provides experts, advisers and consultants on a non-political ideological basis. However, a critical review of the Monthly Lists of Appointments indicates a higher correlation between the experts', advisers' and consultants' countries of nationality and their countries of duty. On the basis of this correlation, we find Czechoslovakian, Soviet, Romanian and Yugoslavian experts, advisers and consultants assigned to predominantly Eastern European aligned countries. British experts, on the other hand, go to most Commonwealth states, while French experts go to Francophone African states. The United States, however, has no definitive policy on assignment of its nationals.[87]

Bilateral donors, on the other hand, tend to concentrate their assistance resources, experts, advisers and consultants into relatively fewer countries. This concentration of aid significantly increases the veto power or authority to foreclose which a donor might have on the development efforts of the recipient country. Multilateral programs, on the other hand, have a distributive effect since development funding reaches more needy and developing states. Disputes are easier to resolve under the multilateral than the bilateral programs. For example, the United Nations OPEX program, Article 11(3), reads: 'The parties hereto recognize that a special international status attaches to the Officers made available to the Government under this Agreement, and that the assistance provided hereunder is in furtherance of the purposes of the organizations. [Article V(2) further stipulates that] any dispute between the organizations and the Government arising out of or relating to this Agreement which cannot be settled by negotiation or other agreed mode of settlement shall be submitted to arbitration at the request of either party to the dispute pursuant to paragraph 3 of this Article [which, in essence, reads that] any dispute to be submitted to arbitration shall be referred to three arbitrators for a decision by a majority of them. Each party to the dispute shall appoint one arbitrator, and the two arbitrators so appointed shall appoint the third, who shall be the chairman.'[88]

To what extent may developing governments assert effective control over bilateral experts, advisers and consultants as far as their work is concerned in the recipient country? The general conditions formulated by the United Nations Secretariat in document ST/AI/232 of 28 November 1975 suggest that, for the purposes of Article VI of the Convention on Privileges and Immunities of the United Nations, 'they are experts on missions for the United Nations. Consultants, experts and contractors, when on business of the United Nations, shall neither seek nor accept instructions on this matter from any Government or from any authority external to the United Nations. They may not engage in any activity that is incompatible with the discharge of their duties with the Organization.'[89]

Developing governments accept bilateral or multilateral experts, advisers and consultants with very little knowledge, if any, of the productivity or the professional quality of their work. The processes of recruiting, screening and testing experts, advisers and consultants are undertaken by the bilateral and multilateral donors themselves. The United Nations Development Program affirms: 'No consultant was put on the Roster without prior endorsement from an officer acquainted with the subject matter, mainly of the Technical Advisory Division, and an officer of the Division of Personnel. However, an *ex post* evaluation is particularly important to ensure that only consultants of high standard are *kept* on the Roster. Therefore, the officer directly responsible for the consultancy, either in the Field Office or at Headquarters, is required to comment on the consultant's mission. While no formal set of criteria has been set up, one should be guided in one's judgement by the fact that the Roster should contain only those consultants who will most likely succeed on a similar mission in another country.'[90]

As a minimum, bilateral and multilateral agreements protect experts, advisers and consultants in a way that leaves a cost-effective or cost-conscious developing country desperate for ways to control the quality and performance standards of these professionals. In assessing the implications of the politics of the *persona non grata*, it is appropriate to ask to what extent does this instrument serve as a substitute for expressing the helplessness developing governments feel in the face of the numerous bilateral and multilateral legal technicalities. Through the invocation of the *persona non grata* or letters of exchange, a desperate recipient expresses displeasure at the acts of a donor. However, neither the invocation of the *persona non grata* argument nor its actual use guarantees a developing government effective control of these professionals. Formally or informally, the dictum that 'beggers cannot choose' applies here. On the contrary, the potential threat posed by the force of the *persona non grata* on experts, advisers and consultants does not affect their value perceptions. For one thing, the invocation of the *persona non grata* argument suggests that at least a developing government is sensitive to criticisms of its organizational irresponsiveness or inflexibility to bilateral and/or multilateral interests. The net effect of these practices on the politics of efficiency is that 'it is unavoidable that standards both of morality and of effectiveness are apt to be radically lowered' in favor of technical expediency.[91] It is difficult to introduce higher standards of efficiency unless both parties adhere to its terms and adopt measures for its institution.

DEVELOPING PROFESSIONALS

There is a contingent in African professionals who meet our criteria of experts, advisers and consultants but who, because of the recruitment practices of major technical assistance programs and governments, are not used well, if at all. The African expert, adviser and consultant is a recent by-product of philanthropy of the 'Big Three' American foundations – Rockefeller, Ford and Carnegie.

The Rockefeller Foundation brought to the United States and Europe a total of 211 African para-professionals between the years 1961 and 1967. This number was distributed among 54 US and European universities and studied for advanced degrees ranging from Hospital Nursing Administration to the History of Ideas.[92]

A great number of these Rockefeller Fellows returned to Africa with the conviction that they possessed the prerequisite skills to excel in their preoccupations.[93] Some of them left from America to enter the universities of Oxford and London to settle into the bloodstream of English scholarship. It is this group of African scholars which has contributed most to the clarification of the philosophical and political foundations of African development and, correctly, ought to be acknowledged as experts in African development.[94]

Twenty of these African professionals formed the African Association of Political Science in 1973, based at the University of Dar-Es-Salaam in Tanzania. Political science is a discipline that, for them, synthesizes many disciplines and branches of learning. Further, it has direct reference to the structure of their societies. Equally important, political science is a direct descendant of the grand human discipline of philosophy, from which all branches of learning and scholarly endeavor claim origin.

The processes by which these African professionals intend to influence development policy are through research and publication, since these are the two trademarks of scholarship. To paraphrase Claude Ake: The aim of the research and publications program is to promote and decolonize knowledge and to facilitate the solution of

problems of particular interest to the peoples of Africa. It is quite
evident that Africans cannot rely on others to research and solve
their problems for them. For instance, a UNESCO source estimates
that at most 1 per cent of the research done in the industrialized
countries is directly relevant to the developing countries. It is
unlikely that other people can research and 'solve' Africa's problems
satisfactorily since they have values and interests that Africans do
not necessarily share. More importantly, unless Africans take it upon
themselves to define their problems and devise solutions for them,
Africa cannot overcome her stagnation, dependence and her exploitation by others.[95]

Ake's declaration of principle has meaning only if, after evaluating
the strengths of research facilities and the cooperation of universities
and governments in Africa, research and publication can increase the
probability that there can truly emerge an African-based research
with direct bearing on a truly African problem. We have no evidence
upon which to evaluate the truthfulness of the assertion that there is
specifically an African problem whose solution lies in a truly African-
based research. Science, Tom Mboya argued, is a universal property
which transcends all cultures and ambivalences.[96] It can be indigenous,
but not truly indigenous.

The Organization for African Unity (OAU) suggests, unlike Claude
Ake, that the scientific problems of the continent can best be
overcome by intensification of cooperative efforts, mobilization of
available scientific skills, acceleration of the processes of its application, and an increase in the awareness among member African
states that improved knowledge about the fundamental contributions
of science to growth is immense.[97]

There are seven areas of research and intensified scientific activity
in Africa, according to the OAU resolutions. The first one (not in
order of priority) is *labor*. Resolution CM/Res. 24 (II) of February
1964 suggests that development without equitable and fair labor
practices is tyrannical and contradicts the foundations of egalitarian
humanism. In subsequent resolutions – CM/Res. 194 (XIII) and
CM/Res. 129 (IX), adopted on 10 September 1967 – the Organization
stressed the importance of labor dispute reconciliation codes that
conform with those of the International Labour Organization. To the
Organization's dismay, no comparable and fair labor codes yet exist
in most member African states.[98]

The second problem area is *finance* which is the backbone of the
development effort in Africa. The historical, political and trade

patterns of most African governments render the making of tariff rates uniform and the removal of trade barriers difficult. Finance, until these preliminary barriers are removed, will continue to flow from Europe and North America. The Organization itself recognized this problem and, instead of advocating drastic moves, encouraged its member states through resolution CM/Res. 179 (XII) of February 1969, to use the International Monetary Fund's Special Drawing Rights (SDRs) scheme and link their development finance to SDR parity rates. The most the Organization promised to do was help its member states to negotiate with the IMF to lengthen the SDR repayment period from the usual 3–5 years to 5–8 years.[99]

The third problem area deals with *industrialization*. Industrialization, according to the Organization's CM/Res. 246 (XVII) and CM/Res. 276 (XIX) of June 1971 and 1972 respectively, is a process which came to Africa from Europe. It is a process which is perfected by the application of selective incentives to attract investments, either locally or externally. In both cases, investment requires an entrepreneurial attitude and the element of risk. The preoccupations of the Organization and its member states are focused on methods of evaluating the existing industrial policies, how they can become competitive and fair to a diverse number of investors, and what role governments should play to guarantee investors against losses or risks which arise from human prejudices.[100]

The industrialization objective is tied to the question of *law*. Resolution CM/Res. 27 (II) determines how the Organization could serve member states, as an impartial party, to coordinate, process and resolve legal disputes that may, in the future, arise between them and foreign investors in Africa. Resolution Cm/Res. 199 (XIII) of September 1969 provided an answer to the effect that, by the terms of inception, the Organization cannot really assume the role of an adjudicative body to the embarrassment of some of its member states. However, should member states and potential investors agree on this proposal, a network of treaties can be concluded and deposited with its Secretariat.[101] By the language of resolution Cm/Dec. 108 (XIV), this issue has, since 1970, been under study by the member states.

The fifth problem area is *food and health*. Resolution CM/Res. 112 (IX), on the 'establishment of Regional Stocks of Food and Grains', adopted in September 1967, expresses a fundamental concern over human survival in a continent of drought and climatic contradictions. In September 1968, by resolution CM/Res. 172 (XI), the issue was sent for further study, for lack of a comprehensive method of how to achieve such a monumental project. The health issue, on the other hand, was referred to the World Health Organization since it is this agency that has traditionally played the health scout role for the African governments. The OAU recommended that duplication of health programs would be wasteful and cost-ineffective since the work of the World Health Organization would be more effective if member states shared with it more relevant data on their health needs.[102]

The sixth problem area deals with *conservation and natural resources*. Resolution CM/Res. 118 (IX) of September 1967 argues that conservation and natural resource utilization policies of African states should be 'guided by common provisions in one Convention in the scientific conservation, preservation and exploitation of all of their natural resources for the benefit of man'.[103] 'Man' is a value-loaded term. Neither the above-mentioned resolution nor any subsequent decisions on conservation and natural resources defines it well. However, this terminological difficulty does not explain why the advocated convention has, to date, not been concluded.[104]

The seventh problem area involves *social affairs*. Historically in Africa, social affairs received little policy recognition since the colonial governments adopted the view that they could not adequately assume the burden of economic civilization while, at the same time, be distracted by human values that have roots in tradition. Resolution CM/Res. 128 (IX) simultaneously provided a definition of both social affairs as a policy area that deals with fundamental human values and the role of government in providing refuge to traditional values that may otherwise succumb to the development pressures. At the same time, the Organization provided its member states with a working definition or use at the forthcoming World Conference on Social Affairs in 1968. By resolution CM/Res. 163 (XI), member states adopted the Organization's recommendations.[105]

African professionals do not have to beg the question by arguing that the contexts and tools of their scholarship can be truly African, since laws of logic do not seem to support that contention. However, they are correct in suggesting that, from the evidence provided in this study, the governments and universities of Africa are not doing enough to sponsor research by African scholars. Much of the research done in Africa is an offshoot of the research of international organizations, foreign governments, multinational corporations, foundations and universities. Associations of African scholars and professional specialists rely essentially on foreign sources for their operating budgets and their research funds. Despite all the talk of African personality and self-reliance, African governments and universities continue to acquiesce in a situation in which foreign expertise, foreign institutions and money control research in Africa.[106]

The argument, if it can be summarized, identifies the structure of knowledge in the world as essentially biased and detrimental to secondary professionals who are not free to enter and utilize it. Since African professionals are denied the exercise of their expertise in solving development problems, they will, for some time, be disenchanted. The argument of expertise will be seen as perpetuating servitude. The knowledge of servitude is discriminatory since it classifies people as unequal, and such knowledge is rooted in the history, politics, economics and sociology of the societies that value it most.

Developing governments seem to have no answer to this problem. Indeed, the answers to the numerous criticisms about how overly dependent developing governments are on foreign experts, advisers and consultants cannot really come from them since, by virtue of the vicious circle of their operations, they cannot produce assertive solutions in a controversy which is grounded in the mythology of technology. Here, again, African governments neither pay experts nor are responsible for their selection. What is important, therefore, is to discover the root of this issue and solve it. The newly formed African International Consulting Consortium may be expected to have a role to play in mitigating against this historical deficiency.

CONCLUSION

This study has raised two fundamental problems. The first proceeds from the observations that nearly every development plan that has been formulated by bilateral and multilateral experts, advisers and consultants has failed. The critique takes experts, advisers and consultants as sacred individuals whose projections and models possess divine precision. This view is both unrealistic and heretical since 'Whoever of you imagines he is wise with this world's wisdom must become a "fool", if he is really to be wise.'[107]

The assessment indicates that bilateral and/or multilateral experts, advisers and consultants perform specialized services which are probably too specialized beyond the comprehension of the developing governments to appreciate fully and/or utilize effectively. The solution of this problem lies in the establishment of vertical and/or horizontal or corresponding or counterpart professionals who possess skills to translate these complex professional exchanges. In accordance with the training requirements of the United Nations under OPEX, for example, it is mandatory for recipient governments to provide national counterparts to the OPEX officers. During the initial stage of the project, these counterparts serve as trainees of the OPEX officers. After a prescribed length of apprenticeship, they assume full control with the OPEX counterparts in the background. This process fulfills the stated requirement of establishing compatible values on a project-by-project basis.

Opportunities to learn from experts, advisers and consultants are few because of the short-term duration of the appointments. However, the misconceptions about such professionals, whether in perceiving them as semi-gods or merely as men or women, are caused by lack of knowledge as to who these people really are and how they do what they do. Learning from them would educate most of us to the insecurities and the frustrations that are typical to their profession. These insecurities and frustrations are increased when they have to operate in a new developing environment where the basic tools of their trade are inadequate and the data or information bases at their disposal woefully inaccurate.

Bilateral and/or multilateral experts, advisers and consultants perform very few data-gathering duties while on administrative assignments. Inevitably, they rely on the recipient governments to perform this rudimentary but essential research task. These professionals also feel comfortable if the client can compile the statistics and, if such a database is sufficient for development purposes, share this information with them. The responsibility for errors of omission or commission thereby become those of the recipient governments themselves. To construct a reliable statistical database, it is essential that developing governments possess the necessary research skills within their indigenous professionals that may complement the work of the experts. The African professionals may be of some use in this context. However, the use of African professionals is likely to involve political sentiments, favorable and/or unfavorable, to development efforts. This problem can be alleviated if African governments can treat their indigenous professionals with the fairness and respect that is due them by virtue of their professional integrity. Failing to do this, African professionals will increasingly be critical toward development policies and the shortsightedness of those who devise these policies. Should this problem persist, developing governments will have to rely on their own source of technical expertise.

To conclude, therefore, developing governments have political mandates which they wish to discharge. They use private experts, advisers and consultants, either local or foreign, with the intention of discharging these mandates. Contrary to the misconceptions that are held about them, these governments cannot, and should not be expected to, abdicate their political mandates to such professionals, whatever inclinations they might have toward de facto power. A clear distinction of delegation based on expertise should be made between experts, advisers and consultants and the user clients themselves. Such a distinction will help clear the myths and contradictions that shroud technical assistance programs, both bilateral and multilateral. This distinction, as a prerequisite, might also assess the genuine scope of technical assistance programs and their power to solve the fundamental humanitarian and egalitarian problems facing the Third World. Directly, the discussion of experts, advisers and consultants relates to how developing governments administer their development objectives and the people who evaluate the performance of these objectives. Broadly, the discussion refers to problems of professional perception and the measurements, standards and techniques that are normally in use by other cultures when looking at other cultures. To use Robert Musil's argument on morality, our discussion hinges upon values: 'Values have been shifted round. Certain questions have been taken out of man's heart. What has been set up for the breeding of high-flying thoughts is a kind of poultry-farm known as philosophy, theology or literature, where in their own way they multiply and increase beyond counting; and that is quite convenient, for faced with such expansion nobody any longer needs to reproach himself with not being able to look after them personally.'[108]

Notes

1. Herbert Spencer, *Social Statistics* (New York: n.p., 1883), pp. 11–13, and Herbert Wallace Schneider, 'Science and social progress: a philosophical introduction to moral science', reprinted from *Archives of Philosophy*, no. 12 (Lancaster, Pa: n.p., 1920).
2. Daniel K. Mbila, 'Development Planning in Cameroon', *Economic Bulletin for Africa* 12 (1976), p. 2. See also, J. F. Rweyemamu, 'Development Planning in the United Republic of Tanzania', *Economic*

Bulletin for Africa 12 (1976). The argument regarding Malawi appears in Kathryn Morton, *Aid and Dependence* (London: Croom Helm, 1975). The technical experts' data she gives is rather exhaustive. These are: 1964 (1,576); 1965 (1,318); 1966 (1,362); 1967 (1,853); 1968 (1,810); 1969 (1,764); 1970 (1,751); 1971 (1,630); and 1972 (1,512). These totals include assisted personnel. See Table 6.4, p. 146, in Morton.

3. H. S. Aynor, *Notes from Africa* (New York: Praeger, 1969), p. 130. A recent view from Tanzania is provided by Rolf E. Vente, *Planning Processes: The East African Case* (Munchen: Weltforum Verlag, 1970); Arnold J. Meltsner, *Policy Analysts in the Bureaucracy* (Berkeley: University of California Press, 1976); Edward H. Althans, 'Consultancy overseas earnings: a case of good advice', *Barclays Review* (London: Barclays Bank Group, February 1978); Gerald E. Caiden, 'International consultants and development administration', *International Review of Administrative Sciences*, Vol. 42 (1976); and Anne Winslow, 'The technical assistance experts', *International Development Review*, Vol. 4 (September 1962).

4. Gerald M. Meier, 'The role of an expert advisory group in a young government', in E. F. Jackson (ed.), *Economic Development in Africa* (London: Basil Blackwell, 1962), p. 197.

5. Morris Davis, *Interpreters for Nigeria: The Third World and International Public Relations* (Urbana, Ill.: University of Illinois Press, 1977), pp. 21–23.

6. Consult the annual reports of the Attorney-General of the United States government submitted to the Congress on the Administration of the Foreign Agents Registration Act of 1938, as amended for the respective years.

7. Bridglal Pachai's references to the Malawi–South Africa relationship is interesting. See his *Malawi: The History of the Nation* (London: Longman, 1973), p. 299.

8. Ibid., p. 300.

9. Ibid., p. 301.

10. Ibid., p. 302.

11. Consult *A New Latin Dictionary*, as revised and enlarged by Charlton T. Lewis and Charles Short (New York: American Book Company, 1907); and also *The Oxford English Dictionary*, Vol. III (D–E) (Oxford: Clarendon Press, 1933).

12. Herbert Goldhamer, *The Adviser* (New York: Elsevier, 1978), p. 139; see also D. P. Simpson, *Cassell's New Latin Dictionary* (New York: Funk & Wagnalis, 1959).

13. Alfred Hunt, *The Management Consultant* (New York: Ronald Press, 1977), p. 6.

14. Goldhamer, op. cit., p. 154.

15. Hendrik S. Houthakker, 'The Breakdown of Bretton Woods', in Werner Sichel (ed.), *Economic Advice and Executive Policy – Recommendations from Past Members* (New York: Praeger, 1978), pp. 45–64.

16. Guy Benveniste, *The Politics of Expertise* (Berkeley: Glendessary Press, 1972), p. 10.

17. The literature clearly rebukes the view that the humanitarian objective

is different from the economic issue. Since the existing structure of ideological cold war politics determines the direction of the humanitarian debate, economics will play a significant role in its resolution. Reference to the UN Charter is recommended; see Wolfgang G. Friedman *et al.*, *Cases and Materials on International Law* (St Paul, Minn.: West Publishing, 1969), pp. 221–2 and 222–5.

18. Critiques of the US postures on the new humanistic foreign policy argue that the human emphasis negates the historical record of US foreign policy in the past. The UN Charter only expresses intention but not the method for achieving this objective. See ibid., p. 48, n. 19.

19. See, for example, Friedmann, Wolfgand G., *et al.*, 'International Law of Co-operation', Chapter 11, in *Cases and Materials on International Law* (St Paul, Minn.: West Publishing, 1969), pp. 1008–162.

20. Gerald M. Meier, 'Development decade in perspective', in Ronald Robinson (ed.), *Developing the Third World: The Experience of the Nineteen-Sixties* (Cambridge: Cambridge University Press, 1971), pp. 18–37.

21. There are numerous arguments on the economics of scale and their relation to regional politics. A good assessment, however, appears in Mekki Mtewa's 'Economics of Southern African detente', *International Review of History and Political Science*, Vol. 14 (1977), pp. 39–54.

22. *The Republic of Plato* (Cornford edition, op. cit.), pp. 56–7.

23. Smith, op. cit.

24. Ibid., pp. 30–31.

25. See Article 1(1) of the General Agreement on Tariffs and Trade adopted at Geneva, 30 October 1947.

26. It was only in 1923 that the United States competed with the trade of other countries in the markets of the world. Its presence at Havana, Cuba, in the winter of 1947–48 at the inaugural meeting of the International Trade Organization and (on its demise) the GATT, prompted the insertion in Article 1(1) of the word 'unconditional'.

27. Gunnar Myrdal, *An International Economy* (New York: Harper Torchbooks, 1969) provides a good argument. For example, in the appendix, he refers to some of the problems facing GATT, one of which is international economic integration. 'International economic integration . . . is at bottom a much broader problem than trade and even than economics. It involves problems of social cohesion and practical international solidarity, and the building up of machinery for accomplishing inter-governmental agreements and large-scale political settlements, as a half-way house to the common decisions on economic policy that may be out of reach for our age' (p. 340). GATT has no administrative authority within its own right. This, in turn, impairs GATT's potential.

28. Niccolo Machiavelli, 'Discourse', in which it is examined whether the language in which Dante, Boccaccio, and Petrarch wrote ought to be called Italian, Tuscan, or Florentine. In Anthony J. Pansini, *Niccolo Machiavelli and the United States of America* (Greenvale, NY: Greenvale Press, 1969), pp. 1350–7.

29. A good reference here is Max Horkheimer in his 'Traditionelle und

kristische Theorie', *Zeitschrift für Sozialforschung*, Year 6, no. 2 (Paris) (1937); 255, where he writes that human beings are products of history not only in their bearing and their way of dressing, their stature as well as their mentality, but also in their manner of seeing and hearing, which cannot be dissociated from the process of social life as it has developed over the millenia. The facts presented to us by our senses are performed in two ways: on the one hand by the historical character of the object perceived, and on the other, by the historical character of the organ of perception.

30. A British Preparatory Study Concerning a Draft Declaration on the Rights and Duties of States in the Panamanian draft, Article 6. See the memorandum submitted by the Secretary-General, UN Doc. A/CN.4/2, 15 December 1948, p. 66.

31. Friedman et al., op. cit., p. 200.

32. Ibid., p. 158. We presume that, as prerequisites, nationalist sentiments are essential to the formulation of a comprehensive method.

33. Karl Marx, *A Contribution to the Critique of Political Economy*, trans. B. I. Stone (Chicago: Charles Kerr, 1904).

34. Real Article 1(2) of the International Covenant on Civil and Political Rights, as Act of General Assembly, Resolution 2200, 21 UN GAOR, supplement 16 (A/6316).

35. Kwame Nkrumah, *Neo-Colonialism: The Last Stage of Imperialism* (London: Thomas Nelson, 1965), pp. x and xv.

36. Delbert A. Snider, *Introduction to International Economics* (Homewood, Ill.: Richard D. Irwin, 1971), pp. 374–8.

37. For a comparative view, see the Report of Proceedings of the Twenty-First Commonwealth Parliamentary Conference, New Delhi, India, October 1975, especially the panel discussion on 'Building a New International Economic Order' (Panel 4), pp. 225–50; and also the comprehensive Report on the Agency's philosophy in *Manpower Employment Development for Economic Growth and Social Justice* (Washington, DC: Office of Labor Affairs, US Department of State, Agency for International Development, April 1975).

38. For the historical argument which cannot be articulated by anyone else, see F. D. Lugard, *The Dual Mandate in British Tropical Africa* (Edinburgh: n.p., 1922). For a good lucid reading, consult David B. Abernethy, 'The impact of foreign aid on African politics: some plausible propositions', conference paper delivered at The African Studies Association meeting, 29 October–1 November 1975, San Francisco.

39. Eugene R. Black, *The Diplomacy of Economic Development* (Cambridge, Mass.: Harvard University Press, 1960).

40. Donald G. Morrison and Hugh M. Stevenson, 'Integration and instability: patterns of African political development', *The American Political Science Review*, Vol. 66 (September 1972), p. 904.

41. The transactional magnitude is not considered here an incorrect measure of the final content and strength of a national development policy. See Karl W. Deutsch, *Nationalism and Social Communication*, 2nd edn (Cambridge, Mass.: MIT Press, 1966).

42. Morrison and Stevenson, op. cit., p. 904.
43. Warren F. Ilchman and Norman Thomas Uphoff, *The Political Economy of Change* (Berkeley: University of California Press, 1971), p. 244.
44. Yehezhel Dror, *Public Policymaking Reexamined* (Scranton, Pa.: Chander Publishing Co., 1968). Consult also Robert S. Friedman, *Professionalism: Expertise and Policymaking* (New York: General Learning Press, 1971) and Leonard D. Goodstein, *Consulting with Human Service Systems* (Manila, Philippines: Addison-Wesley, 1978), especially Chapter 8, pp. 142–61. For a virtuous assessment of consultants, see F. Steele, *Consulting for Organizational Change* (Amherst, Mass.: University of Massachusetts Press, 1975).
45. Morrison and Stevenson's assertion about the instability of values relates only to pluralistic societies. See Harry Echstein, *A Theory of Stable Democracy* (Princeton, NJ: Princeton University Press, 1961).
46. Tamar Golan, *Educating the Bureaucracy in a New Polity* (New York: Columbia University, Teachers College Press, 1968), p. 50.
47. An argument according to Uma Lele's article, 'African Experiences with Rural Development: A Digest Report on the African Rural Development Study', International Bank for Reconstruction and Development Bank Staff Working Paper No. 195 (Washington, DC: n.p., January 1975), pp. 33–34.
48. Margaret Wolfson, *Aid Management in Developing Countries: A Case Study – The Implementation of Three Aid Projects in Tunisia* (Paris: OECD, 1972), p. 45.
49. In footnote 1 to Chapter VII entitled 'Counterpart staff: arrangements with user departments', Margaret Wolfson observes that in Tunisia, both donors and the Tunisian authorities agree that Tunisian engineers are sometimes better qualified than those provided under aid arrangements. In one OEP project, for example, it was recognized that two of the technicians provided under aid arrangements were less qualified than the local staff and they were accordingly withdrawn (ibid., p. 45).
50. Read the Report from the Select Committee on Overseas Aid, HM Government of the United Kingdom, 1971. Do we, or don't we, need a kind of an affirmative declaration on equal employment, remuneration, and opportunity in access? We attribute the strategic view of this to Thomas C. Schelling, *The Strategy of Conflict* (New York: Oxford University Press, 1973), Chapters 3 and 4. For the other view, see A. L. Adu, *The Civil Service in Commonwealth Africa* (London: George Allen & Unwin, 1969), Chapter 11, pp. 189–99.
51. Here, again, Eugene R. Black's study is helpful. Read his *The Diplomacy of Economic Development* (Cambridge, Mass.: Harvard University Press, 1960). A good argument also appears in Vernon McKay, *African Diplomacy – Studies in the Determinants of Foreign Policy* (New York: Praeger, 1966); Doudou Thiam, *The Foreign Policy of African States – ideological Bases. Present Realities and Future Prospects* (New York: Praeger, 1965); and William I. Zartman, *International Relations in the New Africa* (Englewood Cliffs, NJ: Prentice-Hall, 1966).
52. In a form of a survey, the fringe benefits and other entitlements include

family passages to and from station of duty, installation grant and/or outfit allowance, FSSU or similar superannuation scheme or gratuity in lieu, education or children's allowance, subsidized housing at full rent if possible, medical scheme, home leave with passages at fixed intervals, expatriation or special supplementary allowances. Read the Inter-University Council's *Opportunities for Appointments in Overseas Universities* (July 1978), p. 2; and Morton, op. cit., p. 138.

53. British Aided Conditions of Service (BACS) relate only to Malawi and Zambia; BESS has a considerably wider coverage and application, as has OSAS. In the beginning, OSAS was restricted to members of the Overseas Civil Service or to persons who had been appointed on contract in the same way as British Officers.

54. Report from the Select Committee on Overseas Aid, Cmnd 4687 (London, HMSO) presented to Parliament by the Minister for Overseas Development (June 1971), p. 4; p. 6, paragraph 13; p. 10, paragraph 34; and pp. 10–11, paragraph 38. For a supplemental view, read Colonial Document No. 306, 'Reorganization of the Colonial Service', issued by the Colonial Office in 1954.

55. US economic and military assistance to Africa south of the Sahara came to $292,689,000, of which $246,131,000 was economic. In 1977, however, this came to $427,192,000, of which $353,447,000 was economic. The 1979 fiscal year figure reflects some increase. A good summary of the AID programs also appears in *Issue – A Quarterly Journal of Africanist Opinion*, Vol. 8 (Summer/Fall 1978), pp. 1–2.

56. *US Policy Toward Africa* – Administration Official's Review of US Policy Toward Africa with particular references to Southern Africa, 12 May 1978 (Washington, DC: US Government Printing Office, 1978). A good document is the so-called *The National Security Study Memorandum 39 – the Kissinger Study of Southern Africa*, edited and introduced by Mohamed A. El-Khawas and Barry Cohen, prefaced by Edgar Lockwood (Westport, Conn.: Lawrence Hill, 1976), especially Chapter V, pp. 117–39.

57. References to this argument are numerous. For an overview, however, read Daniel J. Boorstin, *The Americans: The Democratic Experience* (New York: Random House, 1973), p. 575.

58. Ibid., pp. 575 and 579.

59. An authoritative reading is the House of Representatives and Senate Committees on International Relations and Foreign Relations, *Legislation on Foreign Relations Through 1977* (Washington, DC: Government Printing Office, 1978); also the section-by-section analysis of the Proposed International Development Cooperation Act of 1978 for the US Senate Committee on Foreign Relations (Washington, DC: Government Printing Office, 1978), pp. 177–8. The declaration that Africa should be like the free zone encourages the participation in it of a diverse number of governments and their respective private individuals. In the case of the United States, its expenditures from 1967 have averaged 4 per cent annually of its budget total. Since 1972, however, these averages have risen. Additionally, by authority of Section 211(d), the US Agency for International Development spends nearly $10 million

within the United States to enable colleges and universities to develop programs for use by developing governments. In addition to this amount, AID also spends nearly $133,849,751 annually on Project Management Assistance by some 50 agencies abroad.

60. The information appears in the *Directory of Institutional Resources* – supported by Section 211(d) – Grants compiled by US Centers of Competence for International Development (Washington, DC: Office of Program and Methodology, Bureau of Technical Assistance, Agency for International Development, January 1975); read also the *Supplement* to the Directory of Resources for Project Management Assistance prepared by Dr Hyde G. Buller under contract with US AID – Office of Development Administration (n.d.) and the *Current Technical Service Contracts and Grants*, 1 April 1977 through 30 September 1977, prepared by US AID, Office of Contract Management (Washington, DC), pp. 79–107. For additional resource listings, consult the *US Private and Voluntary Organizations*, registered with the Advisory Committee on Voluntary Foreign Aid, 19 September 1978; and also the Reports of American Voluntary Agencies Engaged in Overseas Relief and Development, registered with the Advisory Committee on Voluntary Foreign Aid (Washington, DC: AID, 1975).

61. *The World Bank in Africa* (Washington, DC: World Bank, January 1973), pp. 10–11.

62. World Bank, *Annual Report* (Washington, DC: World Bank, 1978), p. 91.

63. Ibid.

64. Ibid., p. 156.

65. Consult the *Operational Summary of Proposed Projects*, vol. 1, no. 1, compiled by the Johns Hopkins University Press (July 1968), pp. 1–83. There are no actual project figures; other than this deficiency, the summary is complete.

66. For a specific count on each of these geographical segments, refer to specific pages for Africa, ibid., pp. 1–26; Southeastern Asia, ibid., pp. 27–49; Europe, Middle Eastern and North Africa, ibid., pp. 50–66; and Latin American and the Caribbean, ibid., pp. 67–83.

67. World Bank, *Annual Report* (1978), op. cit., p. 91.

68. Like the United Nations or the UNDP, the World Bank has twenty-four broadly defined disciplinary categories with relevant sub-fields. Consultancy is with one major grouping or sub-field. There are restrictions, however; see the Bank's *Discipline Codes for Consultants Roster* (Washington, DC: World Bank, June 1978).

69. All this information is provided by the UNDP Recruitment Division at the New York Secretariat by correspondence and personal interview.

70. A critical review of the monthly appointments lists and inferences from the statistics suggest that there is disparity of representation in the categories and tremendous under-utilization of developing nationals who have the capabilities to act as consultants. There is, however, a problem of identifying these persons in the first place and identifying areas of their effective use. No African government, for example, presently operates a registry of its own consultants for possible use by

the international agencies. The trend will remain uncorrected for some time.
71. For a general summary of this argument, read the *Report on the Twenty-Fifth Session of the Economic and Social Council* – Official Records – Supplement No. 13 (New York: United Nations, n.d.), pp. 10, 11, 12, 21–23, 194–5.
72. Agreements concluded between the UNDP and other member states refer to this program as OPAS; those concluded by the UN refer to it as OPEX. OPAS and OPEX are the same operation program.
73. See, for example, the Treaty Series No. 80 (1963), Agreement Between the United Nations and Great Britain of November 1963.
74. TARS, 'Conditions of Service for Operational, Executive, or Administrative Personnel (OPAS), Part D, May 1971, pp. 15–26.
75. Refer to Document No. ST/AI/232, dated 28 November 1975, entitled 'Use of Outside Expertise and Professional Services', pp. 1–2.
76. Read Article 11(2), p. 3, of the United Nations document TARS-X Contract (66).
77. An old, but good, summary appears in the United Nations pamphlet 'Recruitment of Experts for the United Nations Programmes of Technical Co-operation', June 1970, p. 12.
78. *Aid and the Commonwealth* – Report by the Commonwealth Secretariat (London): Commonwealth Secretariat, 1973), p. viii.
79. Ibid., p. 15.
80. Ibid.
81. Ibid., p. 18.
82. Ibid.
83. Ibid., p. 28.
84. Ibid., p. 30.
85. Ibid., p. 32. According to this report, the government of Kenya alone had a total of 2,791 British officers of whom 1,567 were operational executives; 1,377 were in other government jobs. The information is, however, misleading since 831 OSAS and 160 BESS officers are excluded. Ibid., pp. 59–60. It is not surprising that J. R. Nellis, 'Expatriates in the government of Kenya', *Journal of Commonwealth Political Studies*, Vol.11 (November 1973), pp. 251–64, is illuminating. Professor Gary Wasserman believes that it was part of the independence bargain that British officers should continue to serve in the government. Read his 'The independence bargain: Kenya Europeans and the land issue, 1960–1962', *Journal of Commonwealth Political Science*, Vol. 11 (July 1973), pp. 99–120. The problem is not only a Kenyan dilemma; it also affects the Malawians. Read Richard Hodder-Williams, 'Dr Banda's Malawi', *The Journal of Commonwealth and Comparative Politics*, Vol. 12 (March 1974), pp. 91–114. And for a Nigerian assessment, read Omorogbe Nwanwene, 'British Colonial Policy and Localisation: The Nigerian Experience, 1930–1960', *Journal of Commonwealth Political Studies*, Vol. 6 (November 1968), pp. 202–18.
86. TARS-X Contract (66), a UN contract form, p. 3.
87. Section IV – Methods of Application – of *Recruitment of Experts* for the United Nations Programmes of Technical Co-operation states: 'In

each European country the government has a central office to deal with recruitment of personnel for the Technical Assistance Programme. These offices are known generally as "National Committees for Technical Assistance". Their responsibility, so far as recruitment is concerned, is to find, screen and propose suitably qualified experts for specific posts at the request of the United Nations. For its part, the United Nations has set up in Geneva a branch of its Technical Assistance Recruitment Service to cooperate closely with the national committees and to assist them in their task. In the United States and Canada, governments maintain offices which assist the United Nations in its recruiting activities. At the United Nations, recruitment and coordination with these offices are centralized with the North American Recruitment Office. In countries outside Europe and North America, candidates may apply through the Resident Representative of the United Nations Development Programme (p. 13)'.

88. OPEX (66), Standard Agreement on Operational Assistance form, p. 2.

89. Document No. St/AI/232, op. cit., paragraphs A, B(IV), p. 3.

90. Memorandum on 'Use of Consultants Roster', UNDP/PROG/FIELD/67. UNDP/PROG/HQTRS/84, 26 April 1976.

91. Gunnar Myrdal, *The Challenge of World Poverty* (New York: Pantheon Books, 1970), p. 351. Read, for example, the story of the Karachi officials when the Pakistanis implied that they could buy such [expert] services very much cheaper elsewhere, if they instead were given the dollars for free use. The new government repaid the friendly reception their putsch was given by the United States by suppressing that report, among other things (p. 350).

92. *The Rockefeller Foundation Directory of Fellowships and Scholarships, 1917–1970* (New York: Rockefeller Foundation, 1972). Read also Francis X. Sutton and David R. Smock, 'The Ford Foundation and African Studies', *Issue: A Quarterly Journal of Opinion* (Summer/Fall 1976); and Francis Sutton, 'American foundations and public management in developing countries', in Laurence D. Stifel, James S. Coleman, and Joseph E. Black (eds), *Education and Training for Public Sector Management in Developing Countries* (New York: Rockefeller Foundation, 1977), pp. 117–33.

93. Most of these African scholars, before they went abroad, occupied lower and middle range professional jobs. On their return, they occupied promotional jobs and increased their professional stature through assertive programs of research and publications. As testimony, consult *Who's Who in Africa*, Part 6 (London: Africa Journal, 1976), pp. 1027–364.

94. This assertion relates to the ways in which experts, advisers and consultants become socialized into their roles. The evidence suggests that by all standards, these Africans have gone through the required socialization processes.

95. African Association of Political Science *Newsletter*, p. 2, in which Claude Ake provided a preliminary statement on the intended objectives of the association itself on its formation in December 1973, at Dar-Es-

Salaam, Tanzania.

96. What the African professionals say, although not in the same context, is exactly what Tom Mboya said in the late 1960s, and also what Dr Huda, a Pakistani Minister of Finance and Planning for the government of East Pakistan, said in 1968. Mboya lamented the fact that 'the collection of data, analysis of policy-oriented economic information is too technical and, therefore, unintelligible to the staff of the new administrative structures'. Further, he argued that 'research on European agriculture is not directly applicable to tropical agriculture [since] educational systems in all advanced countries have obvious shortcomings in Africa; and economic policies formulated in fully employed economies have little relevance for us. The research we need must be *done here*, not in Paris or London and not at Yale or Harvard'. See Tom Mboya, 'A development strategy for Africa', ECA, Seventh Session, Lagos, Nigeria, 1968, p. 20. Dr Huda, on the other hand, argued that 'availing of the services of "expatriate specialists", apart from their different social and cultural backgrounds that severely limit their ability to function, governments may ensure that these experts are qualified to advise on the problems assigned to them'. 'Planning experience in Pakistan', *The Pakistan Development Journal*, Vol. 8 (1968), p. 340.

97. This argument follows the reasoning of Resolutions CM/Res. 116 (IX) and CM/Res. 171 (XI) that deal with the requirements for research training in Africa. See Vol. 2 (Addis Ababa: Organization of African Unity, 1967 and 1968), pp. 44–45.

98. Presently, African governments subscribe to the ILO agreement. Locally, they are guided by slim views of labor laws that constitute part of the civil and constitutional framework.

99. It is not certain whether the OAU and its member states succeed in seeking the changes in SDR terms. Incidentally, resolution CM/Res. 213 (XIV) relates to the final outcome of the African Development Fund, per resolution CM/Res. 251 (XVII), adopted at Abidjan in November 1972. See Report of the Conference of Plenipotentiaries. Recently, however, Sierra Leone devalued its Leone and, at the same time, linked it to the SDR instead of the traditional British pound. See *Standard Chartered Review* (London: Standard Chartered, November 1978), p. 34.

100. If understood within the broader context of African politics, the United Nations First and Second Development Decade declarations have continued to address this question. The OAU has not come up with helpful answers.

101. This proposal is dead since there were no responses to the OAU's Secretariat questionnaire dealing with this issue.

102. The most significant document here is the Rockefeller Foundation's working papers edited by Ralph W. Cummings, Jr, 'Food Crops in the Low-Income Countries: The State of Present and Expected Agricultural Research and Technology' (New York: Rockefeller Foundation, May 1976), pp. 73–75.

103. Zambia, through its philosophy of humanism, attempted a coherent definition of 'man' and how this cultural entity has first to be understood

in order to evoke or move it to respond to what governments expect from it. Ibid.

104. The ratification of the referenced convention was to be done in two separate installments. To these problems both resolutions CM/Res. 169 (XI) of September 1968, and CM/Res. 245 (XVII) of June 1971 are pertinent.

105. The governments adopted these recommendations because the World Conference on Social Affairs was only a few months away and they wished to minimize embarrassment because of the lack of a definitive framework of what, in their heads, constituted a social affairs policy.

106. African Association of Political Science *Newsletter*, p. 7. Read also *The African Review*, Vol. 5 (1975), pp. 114–26, Appendix I constituting the Communiqué of the Third World Forum Conference held in Karachi, 5–10 January 1975, and Appendix II constituting the Dakar Declaration and Action Programme adopted at the Sixth Special Session of the United Nations in New York, April 1974.

107. I Corinthians I: 19–21. Also read E. Harris Harbison, *The Christian Scholar in the Age of the Reformation* (New York: Charles Scribner's, 1956).

108. See Burton Pike, *Robert Musil: An Introduction to His Word* (Ithaca, NY: Cornell University Press, 1961), p. 127.

6 Strategies for Communicating on Innovative Management with Receptive Individuals in Development Organizations

Larissa A. Grunig and James E. Grunig

Development professionals who use their expertise to develop plans and policies for Third World countries face an inevitable dilemma: they seldom, if ever, are in a position or have the power to implement their own ideas. Professionals can do little more than recommend options for employers or sponsors. Decisions on the implementation of plans and policies are made by politicians and administrators – not by the experts who formulate them. (Mtewa, 1980, pp. 36–7).

According to Esman (1966), modern physical and social techniques, including science and technology, must be carried out by specialized organizations. In Third World countries, however, these specialized organizations 'either do not exist or require radical restructuring' to implement modern practices. He explained:

> Often individual agencies pursue their specialized functional interests in competition with other units. Their energies are often invested in self-service and self-protection more than in the pursuit of program objectives. Each may be the captive of political figures or of interest groups; their rivalry at many levels of government may drain energies away from their operating function and severely compromise the operating programs of the governing elites. (p. 81)

Many other writers have described the limitations of development organizations (e.g. Riggs, 1964, pp. 260–85; Heady, 1984, p. 283; Sharkansky, 1975, pp. 33–40). To be effective in implementing innovation and change, however, these organizations must be characterized by what Frederickson (1980a, 1980b) has called the 'new public administration'.

The new public administration is based on concepts 'designed to enhance the potential for change in the bureaucracy and to further policy changes that increase possibilities for social equity' (1980a, pp. 8–9). These concepts include 'decentralization, devolution, termination, projects, contracts, evaluation, organization development, responsibility expansion, confrontation, and client involvement' (1980a, p. 8).

Development professionals, however, are not without fault themselves. Experts, advisers and consultants, according to Mtewa (1980, p. 162), often confuse and confound the 'development puzzle' rather than help to solve it. As a result, Mtewa recommended that development organizations have an indigenous, counterpart staff to match the expertise of outside professionals. Outside professionals, he added, should have both horizontal linkages with indigenous professionals in similar roles and vertical linkages with 'policy actors in different strata' (pp. 175–6).

Counterpart staffing is an example of an incremental approach to innovation in development organizations, change brought about internally rather than through wholesale replacement of the organization (Dror, 1976, p. 135). Incrementalism may not be a satisfactory solution when drastic changes are needed for the implementation of a plan or policy, but it often will be the only approach that is acceptable either to the sponsor of a development professional or to the Third World government that hosts the professional.

Whereas structural change is the hallmark of a revolutionary approach to development, communication is the essence of an incremental approach. Communication alone seldom is the solution to the problem of development (Grunig, 1969a), but it can be the instrument by which professionals and national leaders who support development can bring about structural change without resort to revolution.

The purpose of this paper, therefore, is to develop a theory of communication that development professionals can use in working with their counterparts in organizations that administer development programs. As will be shown, communication is an essential ingredient in the 'new public administration' that, in turn, is an essential ingredient in the process of development in the Third World.

COMMUNICATION RESEARCH AND DEVELOPMENT PROGRAMS

Finding the solution to the problem of how to bring about innovation in Third World countries has been one of the most important problems in communication research in the last 25 years (Rogers, 1983), although interest in that problem has not been as keen in recent years as it was in the 1960s. Most of the research on innovation in developing countries, however, has attacked the problem as an individual one. That is, researchers have looked for ways to make individual farmers, entrepreneurs or other decision-makers more innovative (e.g. Grunig, 1968b; Rogers with Svenning, 1969).

In the 1970s, however, communication theorists such as Grunig (1971), Rogers (1976) and Diaz Bordenave (1976) began to point out that it was extremely difficult for individuals to innovate in developing countries because of overwhelming social, political, economic and other structural constraints. Grunig (1971) argued, then, that researchers must study the organizations that promote and manage the programs to stimulate innovations because they are part of the socio-political structure of developing countries that must be changed before innovation will be possible. Most managers in development organizations have a vested interest in the status quo and, therefore, only limited interest in changing the basic structure of society (see Grunig, 1978). On the other hand, changes in the structure of society cannot be implemented and administered except by organizations.

Little research has been conducted on innovation in organizations, however. Rogers (1983, p. 366) reviewed over 2,000 research reports on the diffusion of innovations and found that only a 'dozen or so studies' have been done on innovation in organizations. These studies are few in number, but, in addition, must have limited relevance to the incremental process of organizational and individual innovation that is a key communication problem in Third World countries.

The theory developed in this chapter, therefore, is a theory of organizational innovation in developing countries, the kind of innovation that would seem to be the key first step that must be taken to bring about overall development in those countries.

COMMUNICATION AND IMPLEMENTATION OF MANAGEMENT TECHNIQUES

1. Innovative organizations

Research on organizations in developed countries has concentrated on identifying the characteristics of innovative organizations. Most theorists would call such organizations 'organic'. Organizations produce and adopt more innovations when they have dynamic, complex and large-scale environments (Hage and Hull, 1981) and when their structure is decentralized, less stratified and less formalized (Rogers, 1983, p. 357; and Robbins, 1983, pp. 280–283). Yin (1979) characterized the organization likely to innovate as concerned more with product efficiency than bureaucratic self-interest and one that has excess resources, flexibility and creativity.

Innovative organizations also employ more specialists, grant those specialists greater autonomy, and allow them to participate in key organizational decisions (Hage, 1980, pp. 165–187). (For more on the ideology of autonomy, see Brooke, 1984; Korten and Klaus, 1984; and Lynch, 1983.) Hage and Finsterbusch (undated) have recently reviewed this literature and applied its 'contingency approach' to organizational change in developing countries.

Organizational theory, however, requires an overhaul of the structure of organizations to stimulate innovation (Brinkerhoff and Klauss, 1985, p. 146). The more organic an organization's structure, the more likely it would be that an organization such as DPMC could communicate new management ideas to that organization. But, the dilemma is that most of the management ideas to be disseminated consist of techniques to make the organization more organic. Thus, an organization cannot be made more organic unless it is already organic (see, for example, Fairweather, Sanders and Tornatzky, 1974; and Tornatzky *et al.*, 1980). (For more on this organizational dynamic, or 'transformation problem', see Masuch (1985).)

Hage and Finsterbusch (undated, p. 148) proposed forming new organizations as one solution to this dilemma, especially when the technology to be implemented is disparate. After a revolution, socialist governments form an entire set of new organizations or replace the power structure in existing organizations. Such wholesale replacement of organizations seldom is possible in AID host countries, however.

Thus, an organization such as DPMC must apply the incrementalist approach to bring change in development organizations. Organizational change should begin with small changes in management that have maximum possibility of success (Rondinelli, 1985, p. 229; Mandell, 1983, p. 11; and Solomon, 1985, p. 7). Then, a pattern of successful management innovation could incrementally diffuse throughout the organization until it becomes more and more organic in structure and process.

Incremental change begins with key individuals who perceive a problem in the performance of their organizations in achieving development objectives – a 'performance gap' (Hage and Finster-busch, undated, p. 115). In addition, these individuals must have enough political power in the organization to be able to stimulate or implement management changes. And they must be well connected in an organization's communication network so that they can communicate the value of that change to others in the organization (Rogers and Kincaid, 1981, pp. 239–243).

Within a given organization, several individuals in different formal and informal levels of authority may be involved in any one decision. Their personal relationships and relationships with other organizations and with other developers of management technologies complicate the adoption/implementation process (Rondinelli, 1985, p. 228; Blase, 1973, pp. 8–9; Hafsi, Kiggundu and Jorgensen, 1985; Brown & Covey, 1985; Mandell and Bozeman, 1983, p. 37; Solomon, 1985, p. 6; and Solomon *et al.*, 1981, p. 6). Managers also must balance external adaptation against internal efficiency (Mintzberg and McHugh, 1985, p. 196) and key personnel are often subject to turnover (Bardach, 1984, p. 134).

The resulting lack of unity of purpose often inhibits the decision-making process within organizations in developing countries (Saasa, 1985, p. 314; and Kiggundu, 1985, pp. 22–23). See also Lindblom (1972) for his theory of 'disjointed incrementalism', Quinn (1980) for a related discussion of 'logical incrementalism', and Mandell (1983) for 'partialization'.

It is critical in a communication/implementation program, therefore, to be able to identify the characteristics of powerful but innovative individuals, to understand how and with whom they communicate, and to determine how AID linkers can best interact with them.

2. Communication behavior

Administrators of development programs usually think of 'disseminating' or 'diffusing' the results of their research to linkers and receptive individuals in host-country organizations. Instead, they should think of 'communicating' with linkers and receptive individuals about the new ideas or how the ideas generated by the agency can be integrated with new ideas already being considered by the receptive individuals (Grunig, 1978). As Rogers (1983) explained, new ideas frequently are reinvented by the receptive individuals (p. 175) or generated by them within their own organization (p. 334).

Beginning with research in Colombia in the 1960s, Grunig (1969a, 1969b, 1969c, 1971) developed a theory that explains when people communicate, what they communicate about, and why they communicate. His Colombian research showed that communication seldom brought about the changes in individual farmers (both large landowners and peasants) that lead to development. Rather, opportunities brought about by development stimulated farmers to communicate about taking advantage of those opportunities (see also Diaz Bordenave, 1976; and Rogers, 1976). Grunig's theory of communication behavior also explains many of the findings about communication behavior of innovators and early adopters in diffusion research (Myers, 1985; Rogers 1983, pp. 241–270; and Lionberger and Gwin, 1982, pp. 109–150). The theory also explains the conditions under which 'boundary-spanning' individuals in organizations most actively bring information into the organization (Bales, 1984).

Grunig (e.g. Grunig and Hunt, 1984, pp. 147–159) distinguished between active and passive communication behavior. Actively communicating individuals seek information that is relevant to a problem they perceive from many information sources, as do innovators in the diffusion process. Passively communicating individuals 'process' information that comes to them without any effort on their part rather than 'seek' information. Salasin and Cedar (1982), for example, found that preference for information sources more often corresponds to the user's estimate of the ease of using that source than to the amount of information he or she expects to glean from the source (see also Rosenberg, 1967; and Gertsberger and Allen, 1968). Users have no real need for the information but will pay some attention to it if the information is presented simply and repetitiously.

Many programs to 'disseminate' information, for example, often are aimed at passively communicating individuals who seldom use it and retain little of the information they process.

Grunig used three variables – problem recognition, constraint recognition, and level of involvement – to explain when, why and about what active and passive communication behavior takes place. Each of these variables applies to specific situations, such as, for example, poor performance in managing a credit program in a host-country organization. Individuals are more likely both to seek and to process information if they perceive a problem (high problem recognition) and believe they can do something about the problem (low constraint recognition). Perceived or actual involvement with a situation increases their active communication behavior. Low involvement limits information seeking but not information processing, as long as problem recognition is high and constraint recognition low. For example, high-level administrators in host-country organizations may pay attention to (process) information about problems they recognize in their organizations even if they are not personally involved in dealing with the problems, although they may not actively seek the information.

These concepts also apply to linkers. Turner's (1981) research showed that agricultural extension agents were most likely to give information to clients when they recognized clients' problems, did not feel constrained from helping them, and felt involved with them.

The literature on organizations in developing countries supports the usefulness of these concepts for organizations there. Receptive individuals most often communicate about new ideas when they perceive a problem with performance in their organization (Hage and Finsterbusch, undated, p. 115), although they often do not recognize a problem until they hear of an innovation that will solve such a problem (Rogers, 1983, p. 166). Organizational personnel do not communicate about new ideas because of constraints in implementing the ideas that result from the rigid hierarchical structure of their organizations and society or because information-seeking actually may result in punishment (Bozeman and Cole, 1981). Individuals who participate in organizational decisions (are involved) most often communicate about and attempt to solve development problems (Hage and Finsterbusch, undated, pp. 146–7).

Identifying characteristics of receptive individuals and linkers, based on these three concepts, therefore, is critical for a communication/implementation strategy for a development organization.

3. Characteristics of receptive linkers and individuals in host-country organizations

For the incrementalist approach to work, it is essential to identify receptive AID linkers and individuals in host-country organizations: people who recognize development problems, who are not limited in power by constraints, and who participate enough in decision-making to feel involved with those problems (Bardach, 1984, p. 130).

Receptive individuals, based on the characteristics of innovators in diffusion research (Rogers, 1983, pp. 260–61), should have few economic and social constraints, communicate actively both interpersonally with linkers and co-workers, and have cosmopolitan contacts. Weiss (1982) found that change agents, who view externally generated information as credible and useful, tend to be younger and less experienced than other managers. Development research also shows that receptive individuals who recognize development problems work in organizations with a history of change, have lived in other societies, especially the United States, and are professional personnel rather than civil servants (Hage and Finsterbusch, undated, p. 116).

This research suggests that Mintzberg's (1983, pp. 119–39) categories of groups of people in organizations with different amounts of power might be used in future studies to identify receptive individuals. His categories include the chief executive; middle-line managers; skilled, unskilled and professional operators; unskilled and professional support staff; and the analysts of the technostructure. (For more on the potential of the technocrat, see Bruce, 1985.) Development research cited above suggests that receptive individuals will be professional operators, technical analysts or professional support staff.

Further, level in the hierarchy affects the adoption process. Blakely *et al.* (undated) found that administrators champion innovations that are easy to communicate and that receive support from other non-involved staff; lower-level staff, on the other hand, favor innovations that meet the needs and desires of their peers.

Personality variables confound the problem of identifying individuals who are both powerful and receptive to change. Brinkerhoff and Klauss (1985), p. 153) characterized these managers as entrepreneurial and oriented toward their environment. More specifically, they tolerate ambiguity and are patient, prone to taking risks, able to interact with everyone from peasants to top ministry officials, and willing to 'dejargonize' technical language. They suggested strengthening the interpersonal skills of local-level managers to enhance any social development management programs. Bardach (1984, p. 139) echoed this reliance on interpersonal communication. (Mutahaba, in his analysis of Tanzania in 1986, contended that even top executives can be trained to be more receptive to innovation.)

Saasa (1985, pp. 316–17) emphasized the importance of identifying such individuals or groups, usually among the political élite, who wield power and thus can influence policy. He also noted the problem inherent in these dominant decision-makers who depend more on 'withinputs' than 'inputs' in most developing countries. Finally, as Brinkerhoff (1979) pointed out, oganizational structure, procedures and resources constrain most managers.

The constraint variable of the Grunig theory of communication further suggests the value of the political power perspective on organizations (Mintzberg, 1983; Hage and Finsterbusch, undated, Appendix III; and Stohr and Fraser Taylor, 1981), not only for locating receptive individuals but in finding receptive individuals *with power*. It also suggests the value of teaching empowerment strategies (Moss Kanter, 1983; Brinkerhoff, 1979; Rondinelli, 1985, p. 233; and Kiggundu, 1985, p. 26).

Support for the involvement variable comes from research that shows the value of teams in solving problems in development organizations and the participation and two-way communication that teams facilitate (Hage and Finsterbusch, undated, pp. 145–6; Mintzberg and McHugh, 1985, p. 161). Korten (1983, p. 14) sugges-ted coalition building across organizations.

Turning to linkers, research suggests that receptive linkers are oriented to the client agency's problems rather than to those of their own agency, have empathy for clients, and get out of their offices to contact clients (Rogers, 1983, p. 317).

IMPLICATIONS FOR COMMUNICATION STRATEGY

This review suggests that the DPMC and development organizations like it should base their dissemination strategy on the important distinction between passive information processing and active infor-mation seeking. The strategy should go beyond the simple presen-tation of documents to AID and host-country organizations. Without targeting of receptive individuals, documents would be sent to many unreceptive individuals who might passively process the documents but generally not understand and implement the recommended programs.

It is important to target receptive linkers in AID and receptive individuals in host-country organizations, people who communicate actively about management ideas. We have suggested the characteristics of these target people. If an organization provides documents to receptive individuals, they should process the documents passively. After being alerted to the disseminating organization as a useful information source, the receptive individuals should actively seek additional information and should seek additional interpersonal contacts with AID personnel. Then dialogue, conferences and discussion become important.

The strategy should also pay attention to the connection between AID linkers and target individuals in host-country organizations. A strategy should be devised for helping linkers to contact receptive individuals and work with them to adopt, to implement and – at times – to reinvent management techniques. Linkers also should look for receptive individuals who are already using innovative management ideas – generated internally – so the management techniques can be integrated with these ideas already in use. In addition, linkers should communicate with receptive individuals about empowerment techniques that they can use to effect change broadly throughout their organizations.

This review of literature has produced a theory that offers promise for communication strategy of development organizations and for research on innovation in organizations in underdeveloped countries. The organizations that must innovate to bring about change generally are not innovative. Innovative individuals can be found within them, however. Research suggests how to communicate with the individuals to bring about change in the organizations that must bring about change in their environments.

References

BALES, R. W. *Organizational interface: an open systems, contingency approach to boundary-spanning activities*. Paper presented to the International Communication Association, San Francisco, 24–28 May 1984.
BARDACH, E. 'The dissemination of policy research to policymakers'. *Knowledge: Creation, Diffusion, Utilization*, Vol. 6 (1984), pp. 125–44.
BLAKELY, C., MAYER, J., GOTTSCHALK, R., ROITMAN, D., SCHMITT, N., DAVIDSON II, W. AND EMSHOFF, J. *Salient processes*

in the dissemination of social technologies. Final report to the National Science Foundation, undated.

BLASE, M. G. *Institution-Building: A Source Book.* Washington: US Agency for International Development, 1973.

BOZEMAN, B. and COLE, E. 'Scientific and technical information in public management: the role of 'gatekeeping' and channel preference'. *Administration and Society*, Vol. 13 (1981), pp. 479–93.

BRINKERHOFF, D. W. 'Inside public bureaucracy: empowering managers to empower clients'. *Rural Development Participation Review*, Vol. 1 (Summer 1979), pp. 7–9.

BRINKERHOFF, D. W. and KLAUSS, R. 'Managerial roles for social development management'. *Public Administration and Development*, Vol. 5 (1985), pp. 145–56.

BROOKE, M. Z. *Centralization and Autonomy.* London: Holt, Rinehart & Winston, 1984.

BROWN, D. L. and COVEY, J. G. *Development organizations and organization development: implications for the organization development paradigm.* Paper presented to the National Meetings of the Academy of Management, San Diego, August 1985.

BRUCE, D. C. 'Brazilian technocrats and economic development policy'. *Public Administration and Development*, Vol. 5 (1985), pp. 169–75.

CRESHKOFF, A. J. *Approaches to improving implementation management.* Draft report to the US Agency for International Development, Bureau for Science and Technology, Office of Rural and Institutional Development, Washington, DC, 1985.

DIAZ BORDENAVE, J. 'Communication of agricultural innovations in Latin America: the need for new models'. *Communication Research*, Vol. 3 (1976), pp. 135–54.

DROR, Y. 'Strategies for administrative reform'. In Leemans, A. F. (ed.), *The Management of Change in Government.* The Hague: Martinus Nijhoff, 1976, pp. 126–41.

ESMAN, M. J. 'The politics of development administration'. In Montgomery, J. D. & Siffin, W. J. (eds), *Approaches to Development: Politics, Administration and Change.* New York: McGraw–Hill, 1966, pp. 59–112.

FAIRWEATHER, G. W., SANDERS, D. H. and TORNATZKY, L. G. *Creating Change in Mental Health Organizations.* New York: Pergamon Press, 1974.

FREDERICKSON, H. G. *New Public Administration.* University, AL: University of Alabama Press, 1980a.

FREDERICKSON, H. G. 'The lineage of new public administration'. In Bellone, C. J. (ed.), *Organization Theory and the New Public Administration.* Boston: Allyn and Bacon, 1980b. pp. 33–51.

GERTSBERGER, P. G. and ALLEN, T. J. 'Criteria used by research and development engineers in the selection of an information source'. *Journal of Applied Psychology*, Vol. 52 (1968), pp. 272–79.

GRUNIG, J. E. 'Economic decision-making and entrepreneurship among Colombian latifundistas'. *Inter-American Economic Affairs*, Vol. 23 (Summer 1969a), pp. 21–46.

GRUNIG, J. E. 'Information and decision-making in economic, development'. *Journalism Quarterly*, Vol. 46 (1969b), pp. 565–75.

GRUNIG, J. E. 'The minifundio problem in Colombia: development alternatives'. *Inter-American Economic Affairs*, Vol. 23 (Winter 1969c), pp. 3–23.

GRUNIG, J. E. 'Communication and the economic decision-making processes of Colombian peasants'. *Economic Development and Cultural Change*, Vol. 19 (1971), pp. 580–97.

GRUNIG, J. E. 'A general systems theory of communications, poverty, and underdevelopment'. In Casmire, F. L. (ed.), *International and Intercultural Communication*. Washington: University Press of America, (1978), pp. 72–104.

GRUNIG, J. E. & HUNT, T. *Managing Public Relations*. New York: Holt, Rinehart & Winston, 1984.

HAFSI, T., KIGGUNDU, M. N. and JORGENSEN, J. J. *Structural configurations in the strategic apex of state-owned enterprises*. Working paper No. 85-07 for Ecole des Etudes Commerciales (HEC), Université du Montreal, September 1985.

HAGE, J. *Theories of Organizations*. New York: Wiley, 1980.

HAGE, J. and FINSTERBUSCH, K. *Organizational change and development: strategies of institution-building*. Monograph, University of Maryland Department of Sociology, undated.

HAGE, J. and HULL, F. *A typology of organizational niches based on knowledge technology and scale: the implications for innovation and productivity*. Working Paper 1. University of Maryland: Center for the Study of Innovation, Entrepreneurship, and Organization Strategy, 1981.

HEADY, F. *Public Administration: A Comparative Perspective* (3rd edn). New York: Marcel Dekker, 1984.

INGLE, M. D. & RONDINELLI, D. A. 'Assessing the viability of small industry support organizations'. Reprint from *National Development Modern Government*, November/December 1980, pp. 1–8.

JAHNKE, H. E., KIRSCHKE, D. and LAGEMANN, J. 'Impact assessment of international agricultural research centres'. *Agricultural Administration*, Vol. 22 (1986), pp. 175–96.

KIGGUNDU, M. N. *Sociotechnical systems in developing countries: a review and suggestions for future applications*. Draft report, Carlton University School of Business, Ottawa, Ont., October 1985.

KORTEN, D. C. *Learning from USAID field experience: institutional development and the dynamics of the project process*. NASPAA Working Paper No. 7. Washington: National Association of Schools of Public Affairs and Administration, 1983.

KORTEN, D. C. and KLAUS, R. (eds). *People Centred Development*. West Hartford, Conn.: Kumarian Press, 1984.

LEVINE, D. B., RAZAK, V. & KETTERING, M. *Getting the right people on board: an innovative approach for recruiting, assessing, selecting and preparing technical assistance persons and teams*. Draft working paper for the Development Program Management Center and the US Department of Agriculture, Office of International Cooperation and Development, Technical Assistance Division, in cooperation with the US Agency for International Development, Bureau for Science and Technology, Office of Rural and Institutional Development, February 1984.

LINDBLOM, C. E. The science of muddling through. In Thompson, D. E. (ed.), *Politics, Policy and Natural Resources*. New York: The Free Press, 1972.

LIONBERGER, H. F. and GWIN, P. H. *Communication Strategies: A Guide for Agricultural Change Agents*. Danville, Ill.: Interstate Printers & Publishers, 1982.

LYNCH, T. D. (ed.), *Organization Theory and Management*. New York: Marcel Dekker, 1983.

MANDELL, M. B. *Monitoring and evaluating new managerial technologies*. Paper presented to the Technology and Information Policy Program, the Maxwell School, Syracuse University, December 1983.

MANDELL, M. B. and BOZEMAN, B. with LOVELESS, S. *Toward guidelines for conducting R&D on the guidance-system improvement approach*. Paper presented to the Development Project Management Center, TAD/OICD/US Department of Agriculture, May 1983.

MASUCH, M. 'Vicious circles in organizations'. *Administrative Science Quarterly*, Vol. (1925), pp. 14–33.

MINTZBERG, H. *Power In and Around Organizations*. Englewood Cliffs, NJ: Prentice-Hall, 1983.

MINTZBERG, H. and McHUGH, A. 'Strategy formation in an adhocracy'. *Administrative Science Quarterly*, Vol. 30 (1985), pp. 160–97.

MOSS KANTER, R. *The Change Masters*. New York: Simon & Schuster, 1983.

MTEWA, M. *Public Policy and Development Politics*. Washington: University Press of America, 1980.

MUTAHABA, G. 'The training and development of top executives in developing countries: a Tanzanian approach'. *Public Administration and Development*, **6**, pp. 49–59, 1986.

MYERS Jr, R. E. *Communication behaviors of Maryland farmers: an analysis of adopters and nonadopters of innovations to reduce agricultural pollution of the Chesapeake Bay*. MA Thesis, University of Maryland, College Park, 1985.

NEF, J. and DWIVEDI, O. P. 'Training for development management: reflections on social know-how as a scientific and technological system'. *Public Administration and Development*, Vol. 5 (1985), pp. 235–49.

QUINN, J. B. *Strategies for Change: Logical Incrementalism*. Homewood, Ill.: Irwin, 1980.

RIGGS, F. W. *Administration in Developing Countries*. Boston: Houghton Mifflin, 1964.

ROBBINS, S. P. *Organization Theory: The Structure and Design of Organizations*. Englewood Cliffs, NJ: Prentice-Hall, 1983.

ROGERS, E. M. 'Communication and development: the passing of the dominant paradigm'. *Communication Research*, Vol. 3 (1976), pp. 213–40.

ROGERS, E. M. *Diffusion of Innovations* (3rd edn). New York: The Free Press, 1983.

ROGERS, E. M. and KINCAID, D. L. *Communication Networks*. New York: The Free Press, 1981.

ROGERS, E. M. with SVENNING, L. *Modernization Among Peasants: The Impact of Communication.* New York: Holt, Rinehart & Winston, 1969.

RONDINELLI, D. A. 'Development administration and American foreign assistance policy: an assessment of theory and practice in Aid'. *Canadian Journal of Development Studies*, Vol. 2 (1985), pp. 40.

ROSENBERG, V. 'Factors affecting the preferences of industrial personnel for information gathering methods'. *Information Storage and Retrieval*, Vol. 3 (July 1967).

SAASA, O. S. Public policy-making in developing countries: the utility of contemporary decision-making models. *Public Administration and Development*, Vol. 5 (1985), pp. 309–21.

SALASIN, J. and CEDAR, T. *Knowledge transfer in an applied research/service delivery setting* MITRE Technical Report, McLean, Va. August 1982.

SCHEIRER, M. A. and REZMOVIC, E. L. *Measuring the Implementation of Innovations.* Annandale, Va: American Research Institute, July 1982.

SHARKANSKY, I. *Public Administration* (3rd edn). Chicago: Rand McNally, 1975.

SOLOMON, M. J. *An organizational change strategy for developing countries.* Paper presented to the Development Program Management Center and the US Department of Agriculture, Office of International Cooperation and Development, Technical Assistance Division, in cooperation with the US Agency for International Development, Bureau for Science and Technology, Office of Rural and Institutional Development, rev. 1 May 1985.

SOLOMON, M. J., KETTERING, M. H., COUNTRYMAN, P. J. and INGLE, M. D. *Promising approaches to project management improvement.* Paper presented to the Development Program Management Center and the US Department of Agriculture, Office of International Cooperation and Development, Technical Assistance Division, in cooperation with the US Agency for International Development, Bureau for Science and Technology, Office of Rural and Institutional Development, May 1981.

STOHR, W. B. and FRASER TAYLOR, D. R. (eds). *Developing from Above or Below? The Dialectics of Regional Planning in Developing Countries.* Chichester: Wiley, 1981.

TORNATZKY, L. G., FERGUS, E. O., AVELLAR, J. W., FAIRWEATHER, G. W. and FLEISCHER, M. *Innovation and Social Process: A National Experiment in Implementing Social Technology.* New York: Pergamon Press, 1980.

TURNER, D. M. *Communication behavior of linking agents in the Maryland Cooperative Extension Service.* MA Thesis, University of Maryland, College Park, 1981.

WEISS, J. 'Coping with complexity: an experimental study of public policy'. *Journal of Policy Analysis and Management*, Vol. 2 (1982), pp. 66–87.

YIN, R. K. *Changing Urban Bureaucracies: How New Practices Become Routinized.* Lexington, Ma.: Lexington Books, 1979.

7 Southern African Urban Development: Prospects for Involvement of American Institutions

Marvel A. Lang

For a long time American universities have been providing technical assistance to African countries as part of their research and service missions. Much of their assistance has been aimed at addressing rural and agricultural economic development problems, although social benefits have been derived also. As urban population growth rates have increased dramatically over the past decades in underdeveloped countries, the Southern African region has become one of the most rapidly urbanizing regions in the world. For example, for all of Africa the level of urbanization has risen from 14.5 per cent in 1950 to approximately 30 per cent in 1980, and is expected to reach about 43 per cent by the year 2000. The average annual urban growth rate for the continent continues to exceed both the world average and the average for any other single region (Hardiman and Midgley, 1982). On the other hand, while the urban growth rates for the continent are high, most of Africa is much less urbanized than other regions with some countries having less than 10 per cent of their population concentrated in urban areas (Drakakis-Smith, 1987). Hence, the potential for continued urban growth in Africa is extremely high.

This potential for extremely rapid urbanization to occur on the African continent, and the potentially disastrous consequences without planned development intervention strategies, presents an exceptional opportunity for the involvement of American institutions of higher learning as well as for American philanthropic and international assistance organizations. In Southern Africa the potentials for severe consequences of continued rapid urban growth, as well as rapid population growth in general, are higher because of the environmental and ecological conditions and constraints of the region. Thus, without increased capacity to produce food or to continue to import increasing amounts of food, the countries of Southern Africa could face a potentially disastrous situation.

This paper examines some of the critical issues and problems of urban development in Southern Africa, and some perspectives for the involvement of American institutions in providing technical assistance and expertise to address these problems. The aim, therefore, is (1) to discuss some pertinent issues, problems and priorities related to urban development in the Southern African region; and (2) to discuss some perspectives for the future involvement of American institutions in the region.

PERTINENT ISSUES, PROBLEMS AND PRIORITIES

Although many of the development policies and strategies of Southern African countries are still aimed at rural development and food production, urban development problems in these countries have become more acute in recent decades as even larger numbers of people are flocking to the cities (Tolley and Thomas, 1987). While some of this rural-to-urban migration has been stimulated from the positive consequences of rural development, still too much is the consequence of economic distress in the rural areas and the aspirations to a better life in the cities. The larger cities have been the most severely impacted. While most countries have attempted to develop policies to control urbanization by encouraging rural development, their successes in this regard have been limited.

On a recent visit to the Southern African region this author along with several colleagues undertook a preliminary feasibility study to assess the range and scope of urban development problems, issues and priorities in Zambia, Zimbabwe and Botswana. The aim of the study was to obtain from a diverse constituency of African officials, scholars and lay-persons their perspectives on the problems of urban development and appropriate strategies for dealing with those problems. As a result of this preliminary investigation, we were able to derive some general principles about the problems of urbanization in the region, and some specifics about the problems within these countries.

Generally, the attainment of political independence of the countries in Southern Africa during the late 1950s and 1960s was accomplished with high expectations of rapid economic development and growth. It was also generally expected that these expectations would be accomplished with substantial aid and technological assistance from the West (Doherty, 1986). The indigenous populations naturally looked toward the cities as the logical places where these expectations would be initially fulfilled, since they had exhibited the potential for wealth during the colonial era.

These expectations for the most part, however, have not been forthcoming. Instead, the streams of peasants who have flooded the cities have mainly exchanged the squalor of rural poverty for the greater miseries of the shantytowns and squatter settlements of the large cities. With subsidization of food costs and social services in many large cities as a matter of policy, many still find it easier to subsist in the larger cities participating in the informal economy than attempting to persist as self-sufficient rural farmers.

The urban development problems that persist in the Southern African region, especially in the three countries named above, are partially the results of the pre-independence spatial economies. They are also the results of the political administrative and economic adjustments that have occurred or have not occurred since independence. Breese (1966) has pointed out, for example, that the coming of independence inevitably required the creation of new administrative units and the reorganization of business and industry. In some instances these adjustments are still evolving, or they are yet on the verge of being made. These adjustments notwithstanding, migration to the cities, in most instances, is still occurring at much faster rates than the development of the capacity of the cities to employ, house, feed, service and educate. The net result is the inevitable strain and undermining of the already frail structure of public order and the retarding of planned and sustained economic development.

Nevertheless, given these considerations and the fact that the urban population in these countries has begun to outgrow the rural populations, in comparative rates if not in absolute numbers, none of the countries mentioned has established policies and programs to deal strategically with urban development problems. Their philosophy is that the cause of their urban problems is their lack of capacity to address adequately their needs for substantial rural development. Richardson (1984) points out, however, that even in those countries committed to rural development as the key to their development strategy, there are sound reasons for giving some attention to the urban sector. One such reason is that given the size of the rural population, the dynamics of rural population growth, and the existing agricultural, environmental, resource and technological constraints in many countries, they cannot realistically support expanding rural population growth any better than expanding urban population growth. Hence, the key to raising rural living standards is through migration to urban centers thus reducing underemployment in the rural areas. Zambia and Botswana are excellent examples of this situation in Southern Africa.

Consistently throughout Southern Africa, as in other developing regions, the major problems associated with urban development were identified as: poverty, unemployment and underemployment, food shortages and food security, rapid population growth, lack of adequate social services and amenities, inadequate infrastructural development capability, and the lack of institutional and organizational capacity to address those problems adequately. Of course the degree of severity of these problems in the several countries varies as well as the order of their severity.

These problems are further exacerbated by limited and relatively declining foreign exchange, and other limitations which constrain economic growth. Some specific scenarios will further exemplify this thesis. In Zambia, for example, the lack of alternative mineral resources to replace copper as the mainstay of its economy and its major source of foreign exchange has left the country with an enormous foreign debt and with few choices for economic development. The decline in the world market demand for copper has drastically affected employment in mining and in other sectors of Zambia's economy, to the extent that per capita income has been cut in half since 1980 to about US $200 per year. With substantial numbers migrating from rural to urban areas because of the scarcity of food in the rural areas, food shortages in the urban areas have become an increasing reality since the country can no longer afford to import at the levels it did before the recent economic decline.

While Zambia has never invested substantially in rural development, it has also adopted a policy of not investing in urban infrastructural development hoping that this will discourage migration to the urban areas. Though the government is anticipating a restructuring of the economy away from copper production and toward agribusiness, there is little indication that this will alleviate the already overwhelming stresses on Zambia's urban centers. Summarily, Zambia's urban development problems are exemplified by too rapid urban growth, limited capital and resources, critical shortages of trained manpower, and the neglect of planning for urban growth and development in hopes that the urban problems will simply go away.

Botswana, on the other hand, has its own unique urban development problems, and its own perspectives on how to deal with them. While Botswana has accumulated substantial reserve revenues from its diamond and gold mines and its beef industry, it still has substantial development needs and problems. Urbanization is still at a very low level in Botswana, although the potential exists for extremely rapid urban growth in the next decade. Gaborone, the capital city, for example, was only a trading post 20 years ago when Botswana gained independence. Now its population has already grown to close on 100,000.

One of the main problems facing Botswana in its development efforts is how to transform its revenues into productive broader based economic development. The lack of technically trained manpower and the strong traditional pastoral socio-cultural system hinders this process. Hence, the government of Botswana is the main stimulator of economic development. The main need is to involve private enterprise in the economic development process. Meanwhile, there is substantial visible evidence that the urban development problems in Botswana will only multiply in the coming years unless appropriate planning and development measures are taken to ensure viable long-term economic development. Already there are signs of housing shortages, overcrowding in some areas, and signs of sporadic and unplanned development occurring in the fringes of its urban areas.

The lack of private capital for economic investment, the lack of a substantial resource base and illiteracy are problems that also hamper development in Botswana. In addition, employment generation, and organizational and institutional capacity building are problems that also must be dealt with along with urban infrastructural development. There are definite and unique push/pull factors that indicate that Botswana is about to face a transitional stage in its development cycle that will lead to rapid urbanization and substantial development problems unless these factors are addressed properly and expeditiously.

Zimbabwe, on the other hand, enjoys a comparatively higher level of economic development, and its urban development exemplifies the classical model of neo-colonialism and industrial colonial development. For example, in Harare, the capital city, one sees many vestiges of the colonial influences of planned, systematic and centrally controlled development. A relative and comparative abundance of alternative resources notwithstanding, Zimbabwe's higher level of development is also accountable to several other factors which include:

1. the expansion of Western capitalism into its urban centers;
2. a more sophisticated system of urban development management and planning administration;
3. a relatively higher level of economic diversification and self-sufficiency in food production; and
4. the concurrent development of rural production areas, urban service centers, increased agricultural production both by peasant farmers and parastatal commercial farms, and planned diversified industrial development.

Also, upon gaining its independence, Zimbabwe had a substantial advantage in its development process. It has been able to maintain this advantage because of its larger pool of trained and sophisticated manpower, and because of the involvement of a diverse array of non-governmental, parastatal, civic, social, voluntary and church organizations in its development effort. Thus, Zimbabwe's urban development problems are mainly structural rather than fundamental.

Manpower development, unemployment and underemployment are problems that continue to be of major significance in the urban development of Zimbabwe. Although its urban centers have a much higher level of private enterprise and entrepreneurial development, there is still a need for significant infrastructural development to occur. Likewise, decentralization and economic diversification continue to be major priorities as well as increasing and diversifying agricultural production. Despite its more sophisticated level of urban development and overall economic development, Zimbabwe still does not have an established urban development policy. Hence, its urban development suffers from the need for specific directions and systematic planning.

PERSPECTIVES FOR THE INVOLVEMENT OF AMERICAN INSTITUTIONS

Before a discussion can be raised about the specific roles of involvement that American institutions can play in providing technical assistance and expertise to address the problems of urban development in Southern Africa, some fundamental philosophical issues should be addressed. First, it should be understood that many factors affect the regional economic development and the specific development of high-density areas in Southern Africa. These factors can be summarized as:

1. the dependence of the region on South Africa and the potentially volatile situation in that country that adds to the econonmic instability and destabilizing circumstances of the region;
2. the need for increased economic production capacity by the several countries and the various constraints to increasing this capacity;
3. the concomitant need for increasing their import and export capability;
4. the demand for strengthened economic policies and planning institutions;
5. the need for infrastructural development; and
6. the need to create employment opportunities and improve the general living conditions of the masses of urban dwellers.

Secondly, it should be considered whether the involvement of American universities and institutions will be from the perspective of intervention or cooperative assistance. In the past too many efforts toward development assistance in Southern Africa have been seen as intervention strategies whose greatest accomplishments were the promotion of the further economic vitality of a developed country rather than that of the developing country. Thus, many development assistance projects have been ill-conceived, reflected Western bias, were technologically inappropriate, failed to assist with developing the institutional capacity of the African nations, and ultimately left few tangible benefits. There is little desire in the Southern African region for new intervention strategies based on pre-planned agendas with a 'take it or leave it' approach. There is, however, a strong interest in cooperative assistance efforts where there will be substantive involvement of African institutions in the planning, goal-setting, conceptualization and implementation processes of these efforts.

Nevertheless, the myriad of problems and issues facing the Southern African nations in their rapid urbanization holds considerable opportunities for American institutions to become involved in cooperative assistance efforts. For example, American institutions could play a significant role in helping these nations address their problems relating to trained manpower shortages and manpower development by helping them develop significant long-term manpower training programs. Of course such programs should be based on training in appropriate technology that will assist the overall economic development plans and schemes of the particular countries. Just as too many manpower training programs in the US, for example, have trained people for jobs that no longer exist, there is no need to train manpower in Africa for jobs that may never exist.

Likewise, American institutions could play a major role in assisting these countries to address their problems of illiteracy and the development of their education systems. In Zambia, for example, the government has attempted to provide universal free primary education up to seven years, and to provide for the preparation of Zambians to replace expatriates as education personnel. However, few Zambians continue beyond lower primary and relatively fewer go beyond lower secondary levels (Kaplan, 1979). American institutions could provide substantial personnel training, program development assistance, and on a limited basis, personnel to help alleviate this and similar problems throughout the region.

There are other inherent problems where American institutions can bring to bear considerable expertise and experience in providing cooperative assistance to alleviate and ameliorate urban development in Southern Africa. From a philosophic perspective, however, American institutions should be careful to not continue to assume that 'development' and 'economic development' are synonymous. As Mabogunje (1981) points out, there is much development that can occur which does not directly entail economic development, but eventually leads to it, such as technical training and skills development. Further, another assumption that should not be made is that urban growth in the Southern African region, as well as in other developing regions, in and of itself will mean economic growth (Tolley, 1987). It should be noted also that although the urbanization process has the same connotation in the Southern African region as in other regions, it does not necessarily depict the same contextual aspects of morphology, structure and function as in developed regions. For in this region and other developing regions, many living in urbanized areas still live a very rural socio-cultural lifestyle (Hardiman and Midgley, 1982).

What then are the prospects for the involvement of American institutions for providing assistance in addressing the problems of urban development in Southern Africa? The prospects and the opportunities indeed are many and varied. This author's recommendations in this regard would include the following. American institutions should proceed by:

1. determining what particular expertise and resources could be provided;
2. identifying and initiating cooperative relationships with pertinent

constituent institutions, agencies and individuals within the region;
3. formalizing those relationships and linkages by developing collaborative projects and programs; and
4. beginning to serve as advocates to American philanthropic and assistance agencies for increasing and redirecting the focus of American philanthropy and assistance in the region.

References

BREESE, G. *Urbanization in Newly Developing Countries*. Englewood Cliffs, NJ: Prentice-Hall, 1966.

DOHERTY, J. 'Social geography and development in sub-Saharan Africa'. In Eyles, J. (ed.), *Social Geography in International Perspective*. Totowa, NJ: Barnes & Noble, 1986.

DRAKAKIS-SMITH, D. *The Third World City*. New York: Methran, 1987.

HARDIMAN, M. and MIDGLEY, J. *The Social Dimensions of Development: Social Policy and Planning in the Third World*. New York: John Wiley, 1982.

KAPLAN, I. (ed.). *Zambia: A Country Study*. Washington, DC: US Government Printing Office, 1979.

MABOGUNJE, A. L. *The Development Process: A Spatial Perspective*. New York: Holmes & Meier, 1981.

RICHARDSON, H. W. 'National urban development strategies in developing countries. In Ghosh, P. K. (ed.), *Urban Development in the Third World*. Westport, Conn. Greenwood Press, 1984.

TOLLEY, G. S. 'Market failures as bases of urban policies'. In Tolley, G. S. & Thomas, V. (eds), *The Economics of Urbanization and Urban Policies in Developing Countries*. Washington, DC: The World Book, 1987.

Section IV
Communication Policy

8 Global Telecommunication Strategies for Developing Countries

Peter Habermann

INTRODUCTION

Ten years ago, under the impression of greater coherence with the development process in many Third World countries, the International Telecommunication Union (ITU) began to emphasize in earnest telecommunication as an important ingredient for development.

The report by the Maitland Commission[1] was preceded by a detailed study related to various forms of telecommunication applications in Third World nations. Just in time for the World Communication Year in 1983, *Telecommunications for Development* mapped out point by point possible and ongoing applications of telecommunication infrastructures as complimentary to the development process.[2]

In mid 1985 the stage was set for a world-wide initiative to put the development of telecommunication infrastructures on a high priority within global and national policy planning.

In order to understand the development of the resulting global telecommunication strategies it may be useful to look more closely at the interests of some of the major players in this field.

The ITU, although not totally unaffected by the New World Information Order discussions within UNESCO, had demonstrated that within establishing international standards on a technical base, rational discussions between countries belonging to different power-blocks were still possible. In addition, the smaller action radius of UNESCO resulting from the withdrawal of USA support for this agency had left a vacuum, especially within the field of electronic media, which the ITU increasingly tried to fill, improving its chances for substantial increase of operational funds from the United Nations

143

Development Program (UNDP). The ITU seemed to be determined to absorb a good part of communication advisory functions within national governments which traditionally had been the domain of UNESCO.

For the United States, a strong engagement in the growth of global as well as of national telecommucation infrastructures was likely to net a compensation in power for the influence lost as a result of its withdrawal from UNESCO. Furthermore, much of the engagement, especially in training, was seen as seed investment which later on, in all likelihood, would result in a rather stable share for US industries in the global telecommunication market.

For the Far East, and here especially for the Japanese electronic manufacturing industry, the opening of other markets meant a relief from threatened or already implemented import duties imposed by the USA. It was assumed that due consideration of the needs of Third World countries would set the stage during the next decade to propel Far East manufacturers into a direct and successful competition in telecommunication with US-American and Canadian industries.

The developing countries themselves, which experienced a harmful reduction of financial assistance for their communication systems in the wake of the New World Information Order debate, began to see the increased emphasis of international organizations on their rather rudimentary telecommunication infrastructures as a vehicle to boost national communication systems which before had to occupy a rather low priority within their national development plans.

Non-aligned countries under moderate Eastern block influence welcomed the current importance of telecommunication to assert themselves as members in good standing within respective ITU forums. Satellite footprint discussion of the late 70s had conveyed a good amount of power to non-aligned countries. Now their collaboration was necessary to establish agreed upon distribution patterns of orbital slots for the next generation of geostationary satellites. Their consent was also needed to progress with the implementation of multipath high density telecommunication systems. As an exchange for their goodwill and cooperation within the ITU, non-aligned countries usually opted for compensation in kind, e.g. participation and access to the most modern telecommunication exchange systems.

TELECOMMUNICATION INFRASTRUCTURES IN THE CARIBBEAN BASIN

This introductory analysis was meant to explain the fact that, besides its cultural, historical and political diversification, the Caribbean Basin presented during these years a rather unified picture towards the development of telecommunication infrastructures within the respective national confines as well as on a regional basis. It is virtually impossible to trace all activities in this field tabled by country, by communication service, by technology use, and by intended goals. It seems more important to analyze existing similarities in the development of telecommunication infrastructures in order to extrapolate a trend which marks the beginning of a future telecommunications network which is likely to evolve in the Caribbean.

DEVELOPMENT OF THE CARIBBEAN TELECOMMUNICATION SYSTEM

The analysis of major components within the Caribbean telecommunications system shows two unique factors which set this region apart from developments in other parts of the world.

First, historically the need for telecommunication services arises earlier in a region which consists mainly of a number of isolated islands. Obviously, point-to-point telecommunication facilities using long distance channels are infinitely more time efficient than physical message delivery.

Second, the relatively late independence of most Caribbean islands from former colonial powers resulted in a homogeneous development of communication systems throughout the region during the most critical initial stages. It is therefore not surprising that the British company Cable & Wireless has operated throughout the Commonwealth Caribbean for more than one hundred years.

Cable & Wireless provided a network of submarine telegraph cables which linked the English-speaking Caribbean during the first half of this century. This system, parts of which are still fully functional, can be considered as probably the first fixed long-distance interchange system outside military use which provided a classical structure of telegraph multi-point communication.

Technology improvements were implemented rapidly in this region: commercial tropospheric scatter radio links which used the electro-

magnetic reflection capabilities of ionized layers in the troposphere
were implemented in 1959 and extended later on towards the British
Virgin Islands north and into Guyana to the south-east.

Between 1970 and 1980 those tropospheric scatter systems were
replaced by microwave links which connected most of the islands
from the Virgin Islands group to Trinidad. (It is interesting to note
that the technologically ancient tropospheric scatter links with all
their reliability problems were still used in 1980 and 1981 for mixed
communication systems which linked the University of the West
Indies campus in Jamaica, Barbados and Trinidad.[3])

Shortly after most of the English-speaking Caribbean islands had
achieved their independence from the British Commonwealth, Cable
& Wireless linked the Bermudas through the island chain to Guyana
with a high speed multi-channel system for all types of mixed
telecommunication traffic. The major part of this link consisted of a
submarine coaxial cable.

In 1975 the Caribbean Island Basin was the stage for a new
telecommunication record: a microwave link was established between
Tortola to Saaba bridging over one hundred miles and constituting
the world's longest over-water microwave service link. Two years
later this microwave system was connected to the rest of the world
through satellite earth stations which were located in Barbados,
Jamaica and Trinidad. Those communication links were supplemented by the still existing submarine coaxial cables.

The last five years have seen a rapid development of telecommunication technology which was by and large characterized by a transformation from analogue to digital systems. Digital systems have a number
of advantages which makes them especially applicable in situations
where a high degree of reliability with a minimum of service and
maintenance is required.

Cable & Wireless has planned to transform the existing analogue
networks based on cable or microwave links into an interactive multi-
channel digital system by 1990.

While Cable & Wireless digital replacements will follow the old
microwave connections from Tortola toward the south to Trinidad –
a route which mainly connects former British colonies – American
companies like All American Cable and Radio together with ITT
take care of a similar development from St Thomas to Puerto Rico,
the Dominican Republic and from there onward to the United States.

Combined, the various systems in their final stages will be capable
of carrying nearly two thousand simultaneous telephone channels and

will also carry voice-text, facsimile data and video telecommunication services. It should be noted that this traffic mix cannot be achieved with analogue telecommunication carriers.

Projecting new developments in high density traffic technology it is foreseeable that this link will ultimately carry almost ten thousand simultaneous telephone channels which should in all likelihood match actual needs well into the twenty-first century.[4]

Using the Caribbean Basin as an experimental ground, Cable & Wireless is currently working on a trans-Caribbean fiber optic cable project which is supposed to link Columbia, the Dominican Republic and Jamaica to the USA. Once this link is operational and the still existing problems like switch and exchange reliability are improved, it is planned to connect this cable to the Eastern Caribbean via digital microwave links, once again instituting a Caribbean Basin telecommunication system with a high degree of sophistication and state-of-the-art technologies.

At this point it is interesting to note the difference in the corporate policy of Cable & Wireless and that of similar US companies. While the recent decade of telecommunication technology development showed a high degree of specialization within the American telecommunication industry, British Cable and Wireless maintained its generalist approach to regional telecommunication using the most appropriate medium for any of the small and larger hops between the Caribbean Islands and retaining high density traditional coaxial cable configurations for its link to the continental USA and into Latin America. (However, there are indications that the submarine cable system in its traditional configuration has outlived its utility.)

Another characteristic seems typical for the telecommunication development in the Caribbean Basin: available carriers are used more and more by non-governmental institutions to transmit package data, on-line reservation for airlines and a variety of leased line services. On top, most of the systems within the Caribbean Basin based on cable, microwave or satellite hops carry television transmission and other broadcasting signals. Given its predominance which was retained after the end of direct colonial rule within the English-speaking Caribbean during the 1960s, Cable & Wireless has maintained a high degree of visibility featuring offices in most of the former British colonies.

ORGANIZATIONAL DEVELOPMENT

The availability of a technology which allows for a relatively reliable transmission of all kinds of traffic between the Caribbean Islands had already led to a rather rapid development of organizations and institutions which took advantage of those systems.

The Caribbean Broadcasting Union, which since its inception had decried its dependence on American television programs, has started to use point-to-point microwave links for a daily news exchange between Barbados, Trinidad and Jamaica. Furthermore, the Caribbean Broadcasting Union currently plans to establish a Caribbean radio service which could be established via satellite connections either directly to the end users or using the FM side bands of local radio transmitters. This radio service in all likelihood would compete rather effectively with Radio Antigua, a German sponsored regional AM station which also doubles as the Caribbean shortwave relay of Radio Deutsche Welle of the Federal Republic of Germany.

The availability of high density telecommunication channels between various islands in the Caribbean has prompted a variety of private companies and semi-governmental organizations to establish a regular and ever increasing use of those facilities. Those companies formed the Caribbean Association of National Telecommunication Organization (CANTO) late in 1984. CANTO is a group of entities interested or involved in telecommunication which strive to lead in the formation of a regional Caribbean telecommunications policy. At the end of 1986 most of the larger Caribbean Islands – Antigua, Barbados, Bahamas, Curaçao, Grenada, Jamaica, Puerto Rico, Trinidad, Tobago – and Belize and Guyana, have joined the association.

Historically, most of those entities were carefully coached by Cable & Wireless and did not have any independent input into the development of the Caribbean telecommunication system. With the emergence of competitively operated, privately owned telecommunication services, CANTO seized the opportunity to influence directly the future development and current regulations of the interactive telecommunication services offered in the Caribbean.

One has to consider the colonial history of the region and the role of the British owned telecommunication systems to understand fully the potential impact of CANTO. While Cable & Wireless and to a certain degree ITT viewed the Caribbean Basin as either a transition point for transatlantic telecommunication traffic, or as experimental grounds for new telecommunication technologies without too much consideration of existing regional priorities, the Caribbean Association of National Telecommunication Organizations with its governmental support may be in a position to change service priorities established by international carriers. The governmental connection of most of the institutions belonging to CANTO makes it very likely that regulatory issues can be and will be tied to regionally defined telecommunication service priorities which may or may not coincide with the priorities of international carriers.[5]

Also, other state owned or privately owned telecommunication operators are starting to request pieces of the largely increased telecommunication pie: the Puerto Rico Telephone Company (PRTC) has already requested the Federal Communication Commission to permit the establishment of an earth station on Puerto Rico to serve off-island points. Previously, this service was exclusively provided by ITT and All American Cable and Radio. If this request is granted it would mean the end of the ITT dominance controlling the telecommunications traffic from and into the United States main continent.

The American industry together with the State Department is looking at the diversification attempts with a friendly eye. After all, a user-controlled regional telecommunication market will inevitably lead to competitive duplication of systems with ensuing hardware requests. Already the US Telecommunication Training Institute (USTTI) reports that it received 133 applications from Caribbean telecommunication institutions. AT&T International, Bell South Corporation, Motorola Inc., and the Harris Corporation have already provided first training events for telecommunication technicians at the College of the Virgin Islands, Maw Campus in St Thomas. Not surprisingly the training schedule was heavily telephone system biased and included among others an introduction into cellular, radio, telephone and trunk systems. Given the short geographical distance from the Caribbean Basin to the US mainland it is safe to predict that US strategies in opening sizeable telecommunication markets in lesser developed countries will be met with a high degree of success.

SUMMARY

The development of telecommunication infrastructures within a global context does not only occur alongside technological improvements. The inherent flexibility of point-to-point telecommunication traffic and its basic insensitivity toward the content of this traffic propels telecommunication into a position where it successfully replaces physical infrastructures, especially in developing countries.

Unlike railway tracks and roads, however, communication infrastructures by their very nature stretch beyond national confines.

In industrialized countries the assumption was confirmed that telecommunication is most economically viable if it is regulated only within a technical framework in order to secure compatibility of interfaces between various systems.

On the other hand, given the instrumental character of telecommunication infrastructures for the development process, in Third World countries the growth of point-to-point communication links will have to follow national developmental priorities which inevitably require planning and regulation beyond technical standard setting. Of course, the flexible character of telecommunication is likely to hide the clashes between the free market interests of western industrialized countries and the need for planned priorities in Third World nations. Relatively remote from public attention and the glamour of mass media it is also very unlikely that the growth of telecommunication infrastructures will ever lead to a similar debate which is characteristic for the information flow issue. However, given in-built access rules and limitations, point-to-point mediated communication is infinitely more sensible toward one-sided and imbalanced influence than mass media. The unopposed rule of Cable & Wireless and ITT in the Caribbean telecommunication scene for more than a hundred years shows clearly that the structural issues of global telecommunication have not reached a point where critical discussion is considered necessary.

In a sense, telecommunication still awaits its wake-up call.

Notes

1. Maitland, D. (ed.). *The Missing Link: Report of the Independent Commission for World-Wide Telecommunications Development*, Geneva, International Telecommunication Union, December 1984.
2. ITU, *Telecommunications for Development*, Geneva, 1983.
3. Lalor, G. C. *Caribbean Regional Communication Service Study*, Jamaica, 1982.
4. *Caribecom News*, Vol. 1, No. 1 (December 1986), p. 14,
5. Noguera, Philipe. 'Caribbean Association of National Telecommunication Organizations: its role in Caribbean telecommunications development'. In *Caribecom*, Vol. 1, No. 1 (December 1986), p. 16.

9 A Model for Telecommunication Development in Africa

John James Haule

The world's traditional division of poor and rich nations is being exacerbated by its current transformation into information and technology 'haves' and information and technology 'have nots', as a result of advances in telecommunications services and techniques of the last six years.

The gap between Africa and the industrialized countries in telecommunications has been documented in the *Report of the Independent Commission for World-Wide Telecommunications Development* (the Maitland Commission) which urges African and other Third World countries to give a higher priority than before to investment in telecommunications. The First World Telecommunications Development Conference which met in Arusha, Tanzania, in May 1985 established a global target of bringing all mankind within easy reach of a telephone by the early part of the next century.

This is because a modernized national telephone system is now recognized as vital to economic and social development of all nations. The Maitland Commission (1984) stressed that: 'No development programme of any country should be regarded as balanced, properly integrated or likely to be effective unless it includes a full and appropriate role for telecommunications, and accords a corresponding priority to the improvement and expansion of telecommunications' (p. 11). The presence and use of telephone services helps to promote productive business activities, which in turn increases the overall level of economic development (Bebee and Gilling, 1976; Hardy, 1980; Jonscher, 1981; Marsh, 1976; Shapiro, 1976).

But as Saunders (1983) and Hudson (1984) have noted, despite some of the clear benefits of an efficient telephone system, many nations fail to invest significantly in their telecommunications infrastructure, even though there may be a great demand for a telephone service.

The number of telephone lines in Africa – whose population is expected to increase to 900 million by the year 2000 – is the lowest of any region in the world. In 1985, when Africa's population was estimated at 520 million, there were about 3.3 million main lines with a corresponding telephone density of 0.63 main lines per 100 inhabitants. Less than one per cent of the continent's population whose annual growth rate is approximately 5.5 per cent have access to a telephone, and rural areas in particular remain poorly served. In contrast the telephone density in Asia is 2.8 per 100 inhabitants, in Latin America 5.2 per 100, and over 60 in Europe and North America (Norton, 1987).

According to the World Bank, African telecommunications entities are unable to meet the demand for telephone services because they lack adequate skilled manpower in management and finance – and in the engineering field as well. At present, plans for telecommunications development in Africa are dominated by engineering thinking. Consequently, a lot of decisions on planning and development are made on the basis of available engineering information without knowing the financial implications. The key to telecommunications development in Africa is the improvement of the management and financial systems of the national telecommunications entities, and an appropriate increase of trained engineering manpower (Lomax, 1988).

Sullivan (1988) has argued that developing countries should analyze the growth of their information sector in order to provide an indication of their future telecommunications needs and make appropriate investments. The information sector includes all jobs which entail the creation, processing and transmission of information, and its storage. This sector plays a major role in economic development both as the sender and receiver of information flows, and the prime user of telecommunication channels.

Sullivan researched the growth of the information sector and its impact on the development of telecommunications in Ireland and South Korea. He found that rapid industrialization in the 1970s in both countries led to a telecommunications bottleneck as the demand for telephone services for business and residential use rose sharply. The telecommunications entities were increasingly unable to meet the growing demand for service, and governments in both countries were forced to invest much more than anticipated in new telecommunications projects.

Similarly, one could study the growth of the information sector in Africa in order to help planners gauge the continent's future telecommunications needs. But there has always been a vast demand for telephone services in Africa, and this will continue to grow as Africa tries to grapple with its worsening multi-faceted economic woes which include a severe food shortage.

The World Bank and International Development Association have spent about $766.9 million for telecommunications projects through 1986. The total investment in the various projects was $2.14 billion. Between 1985 and 1987, the World Bank was either providing funding or was considering to fund projects in Ethiopia, Kenya, Ivory Coast, Senegal, Burundi and Morocco. According to the World Bank, the overall macroeconomic reality in Africa will require most countries to concentrate in the near and medium term on rehabilitation and on improving internal operating efficiency, and postpone all but the most pressing new investments (Norton, 1987).

The African Development Bank has financed 27 African projects in 21 countries for more than $251 million since it was created in 1972. Under the United Nations Programme of Action for African Economic Recovery and Development, the Bank was planning to lend some $8 billion from 1987 to 1991. If telecommunications continues to receive its normal seven per cent share, some $560 million would be invested in telecommunications projects. The funds will be provided to support projects mainly for repairing existing installations, improving the financial management of operating entities, establishment of the urban trunk and international networks, and rural telecommunications (Norton, 1987).

African telecommunications services are provided by state monopolies. In a majority of African countries, as in the rest of the world, these monopolies are responsible for both postal and telecommunications matters as either government departments or independent public entities. In 16 countries, telecommunications services are provided by a separate telecommunications entity. Many African countries have separated their domestic and international operations, and often, it is the international side that generates revenue which subsidizes various non-telecommunications services of the government's social welfare programs. This cross-subsidization is a major cause of the inability of the African countries to self-generate funds for expansion of the networks (Norton, 1987).

Much of the equipment presently in place in Africa is of the crossbar type with some semi-electronic equipment. Since 1984, most of the equipment installed for international switching centers in Africa has been of the digital type. The transmission systems in Africa consist primarily of equipment installed before 1975. But the Pan African Telecommunications Network (PANAFTEL) has recommended that its member states replace this equipment with modern, low-power, analog and digital microwave equipment that consumes less power as soon as possible. It also notes that fiber optic transmission cables may be appropriate for Africa in the late 1990s.

With very little existing installed equipment in most African countries, the difficulty of integrating old and new communications technologies should present less serious problems than in the developed countries with large existing networks.

African telecommunications experts are advocating leapfrogging into using the latest technologies. Studies point out that the convergence of computers and telecommunications technologies have brought the price per line for small digital exchanges within the reach of Africa's small telecommunications entities.

Under the guidance of the International Telecommunication Union (ITU), a specialized agency of the United Nations, a new project examining the overall needs of Africa was launched in 1987. The ITU is responsible for the planning, coordination, regulation and standardization of telecommunication world-wide. The Africa project is known as the Regional African Satellite Communications system (RASCOM). This feasibility study, which is expected to be completed in about a year, will examine an appropriate mixture of space and terrestrial technology to meet the anticipated needs of Africa through the early part of the next century. In addition to the RASCOM project, other African countries are studying their own needs for satellite communications independently.

There is a shortage of skilled engineers in African countries to handle the more complex engineering tasks in the telecommunications entities. But much more serious throughout the continent is the lack of qualified and skilled managerial and financial professionals. This drawback applies to telecommunications and other sectors of national economies on the continent as well. There is a scarcity of accountants, chief accountants, financial analysts and financial directors, and other professionals qualified in cash-management functions. There is a vital need for the African state telecommunications monopolies to harmonize tariffs and international accounting procedures.

Some African countries have hired international consultants or equipment manufacturers to provide partial or complete engineering services. Such arrangements, despite their excessive cost, could not be of long-lasting value unless reasonably qualified nationals are available to understudy the international experts and to continue the work subsequently.

The availability of adequate and experienced manpower resources would enhance the development of telecommunications networks in Africa in two main areas. First, it would help make decision-makers in the telecommunications entities better informed of telecommunications innovations in other parts of the world. As Lerner (1976) argued, in order for less developed countries to turn innovations made anywhere in the world to their own use, they will have to begin by improving their information technology.

> This would bring the less-developed countries into the world-wide information network and give them early access to innovations that are being discussed and debated anywhere . . . . By acquiring early access to the world-wide information network, less developed countries would participate in the working out of innovations before they even reach the planning and executing phases. They would thus be in the hitherto unattainable position of being able to decide, in the context of their own needs and resources, whether they wish to adopt or adapt or even improve any innovation being proposed anywhere in the world.   (p. 329)

Second, increased manpower would enable African countries to play a more active role in the activities of the ITU than they have done so far. This includes better attendance and participation of African countries at ITU's major World Administrative Radio Conferences (WARCs) such as the High Frequency, Mobile and Space WARCs, as well as numerous sub-group meetings of the ITU's two consultative committees: the International Radio Consultative Committee (CCIR) and the International Telegraph and Telephone Consultative Committee (CCITT). The CCIR studies and issues recommendations on technical and operating questions relating specifically to broadcasting, radio spectrum allocation, and the geostationary orbit. The CCITT is responsible for developing the international standards for telecom equipment network interconnection, service definitions and associated tariff principles.

But at their present level of development, African states not only lack the expertise necessary to participate effectively at a high technical level, they also lack the finances needed for participation. They are unable to send delegations to the many study group meetings always in progress, 'nor can they spare the competent personnel at home to follow developments in the two committees' (Codding and Rutkowski, 1982, p. 104). The two researchers add: 'The situation has created a vicious circle: without participation by developing countries, the study groups and committees have found it difficult to identify satisfactorily questions and problems of specific concern to these countries' (p. 104).

The situation has also led to expressions of dissatisfaction by developing countries against work programs and decision-making processes of both the CCIR and CCITT. One criticism is that the subjects considered by the two consultative committees are disproportionately those which primarily concern the developed states. It is also alleged that committee recommendations pay insufficient attention to the interests of the developing countries. Often recommendations simply register the *de facto* state of affairs in industrialized countries (Jacobson, 1971, pp. 60–1).

In order to respond effectively to these criticisms, the ITU must demonstrate a capability of meeting the rapidly changing technological and commercial needs of the advanced countries, as well as show that it has the ability to secure a transfer of telecommunications resources to developing countries. Efforts towards achieving the latter are already underway. In response to a recommendation of the Maitland Commission, the ITU formally established a Center for Telecommunications Development in July 1985. The Center, composed of a Development Policy Unit, a Telecommunications Development Service, and an Operations Support Group, has as its general mandate, 'the strengthening of ITU advisory services and the delivery of appropriate technical assistance to developing countries' (*Telecommunications Journal*, 1986, pp. 67–8).[1]

As Africa's national economics begin to grow in the unknown future, bringing with it further expansion of the demand for business information which in turn should exert pressure on telecommunications services, African telecommunications managers will be forced to display better efficiency in developing and maintaining their developing networks.

Practical steps African countries could take in efforts to better take advantage of rapidly changing telecommunications technology around the globe include the positioning of telecommunications attachés at selected embassies in the industrialized world, and also in newly industrialized countries such as South Korea, Ireland and Taiwan. African states could learn from the US telecommunications attaché program which was launched in 1986 at 14 key embassy posts, and is geared to train embassy specialists in: '(a) understanding the issues; (b) assessing the host country perspectives and concerns; (c) effectively representing and advocating US objectives' (Dougan, 1987, p. 8).

Another positive step would be for African countries to try to send technical delegations to special events such as the World Telecommunication Exhibitions in Geneva which are sponsored by the ITU. Telecom 87, the fifth in the series of the exhibitions which was held in October 1987 attracted some 900 exhibitors from more than 70 countries. According to the Secretary General of the ITU, Richard Butler, since its inception in 1971, this quadrennial event 'has become recognized as the focal point for presenting telecommunications developments to the world community . . . (Butler, 1987, p. 29).

A model for telecommunication development in Africa is two-pronged: it calls for a vast improvement in the management and financial viability of the telecommunication entities, as they try to adapt technological innovations around the globe in the development of their networks.

Note

1. 'Establishment of the Centre for Telecommunication Development', *Telecommunications Journal*, Vol. 53, No. 2 (1986), p. 67–8.

References

BEBEE, E. L. and GILLING, E. T. 'Telecommunications and economic development: a model for planning and policy making'. *Telecommunications Journal*, Vol. 43, No. 8 (August 1976).
BUTLER, R. E. 'Telecom 87 – The ITU in a changing world'. *Telecommunications*, Vol. 29 (August 1987).
CODDING, G. A. and RUTKOWSKI, A. M. *The International Telecommunication Union in a Changing World*. Dedham, Mass.: Artech House, 1982.
DOUGAN, LADY D. (Coordinator and Director, Bureau of International Communications and Information Policy, US Department of State). Statement before the Subcommittee on Communications, Committee on Commerce, Science and Transportation, United States, 26 February 1987, p. 8.
HARDY, A. P. 'The role of the telephone in economic development'. *Telecommunications Policy*, Vol. 4, No. 4 (December 1980).

HUDSON, H. *When Telephones Reach the Village*. Norwood, NJ: Ablex Publishing, 1984.

JACOBSON, H. K. 'International institutions for telecommunications: the ITU's role'. In McWhinney, E. (ed.), *The International Law of Communications*. Dobbs Ferry, NY: Oceania Publications, 1971. pp. 60–61.

JONSCHER, C. 'The economic role of telecommunications'. In Moss, M. (ed.), *Telecommunications and Productivity*. Reading, Mass.: Addison-Wesley, 1981.

LERNER, D. and SCHRAMM, W. (eds). *Communication and Change: The Last Ten Years – and the Next*. Honolulu: The University Press of Hawaii, 1976.

LOMAX, D. Principal Telecommunications Officer, The World Bank, Washington, DC. Interview with this author, September 1988.

MAITLAND, D. *The Missing Link: Report of the Independent Commission for World-Wide Telecommunications Development*. Geneva: International Telecommunication Union, p. 11, December 1984.

MARSH, D. 'Telecommunications as a factor in the economic development of a country'. *IEEE Transactions on Communications*, Com-24, 716-22, 1976.

NORTON, C. (Deputy Director, Office of Diplomatic and Public Initiatives, Bureau of International Communications and Information Policy, US Department of State). Presentation made at US Department of Commerce conference on International Telecommunications Market Opportunities. Arlington, VA., March 1987.

SAUNDERS, R. J., WARFORD, J. J. and WELLENIUS, B. *Telecommunications and Economic Development*. Baltimore, Md: Johns Hopkins University Press, 1983.

SHAPIRO, P. 'Telecommunications and industrial development'. *IEEE Transactions on Communications*, Com-24, 305-10, 1976.

SULLIVAN, C. 'Overcoming the telecommunications bottleneck: the effect of information sector growth on telecommunication development policies in Ireland and South Korea'. IEEE Transactions on Communications, June 1988.

10 Information Markets, Telecommunications and China's Future

William B. Crawford

ABSTRACT

This article focuses on the role of information markets and telecommunications in China's development. Development is viewed as the process of gaining greater access to economic benefits, and information markets are seen as an important key to development, though they also represent the danger of 'information dependency' for developing countries like China.

A key role in fostering development while avoiding dependency is played by China's pool of 'brain-intensive' workers. Brain-intensive workers are those able to harness the flood of available information and make it useful both for developing information-intensive specialties and for raising the general level of knowledge among the populace.

As the volume of available information increases, harnessing it becomes more challenging, and talent adequate to meet this challenge becomes proportionately rarer, not only in China but globally as well. Through telecommunications technology, however, this talent can be utilized at levels never before imaginable. Possession of a significant share of this talent enhances China's ability to enter global information markets as full participants rather than dependents.

Three models of development – the mechanical, the organic and the cybernetic – are briefly compared within the context of information and development. The cybernetic model is chosen as the most appropriate for guiding China's efforts, since, despite its being the most complex and most demanding in terms of implementation, it offers the most complete treatment available of factors which need to be taken into account for monitoring a system and effectively transforming it.

CHINA ON THE CUSP

China is on the cusp as global information markets develop around her.

Information markets – the generation, transmission, storage, processing, and economic use of information – increasingly hold the key to development. They also embody the threat of a new kind of dependency between developing and developed countries.

Whether information markets present an opportunity or pose a threat depends upon the ability of the developing country to balance its dependency on foreign information and equipment with its own control mechanisms, its capability to 'digest' information for economic ends, and the extent to which it can participate as seller as well as buyer in the information markets.

Non-participation in these information markets does not seem a viable option in a world where survival requires development. Not only are there specific requirements for information – population trends, factors affecting harvests, etc. – there are general requirements for information as well. These general requirements are epitomized by Daniel Lerner's concept of empathy.[1] Empathy, which Lerner used in his studies of development, entails the ability of humans to look beyond the confines of their current state of mind to the experience of others and to possible future states. Information is vital in cultivating empathy.

In recent decades a concerted effort to cultivate certain kinds of empathy has taken place in China. Godwin Chu writes:

> All the Maoist campaigns, whether the Great Leap Forward, the Cultural Revolution or the criticism of Confucius – deliver one message loud and clear: man hs been locked in by traditions, which must be done away with if development is to move forward apace. But the method advocated by Mao to bring about the downfall of traditions is radically different from Western theory and practices. In the Maoist approach, it is not the individual man, but the collective man, that is considered capable of breaking the shackles of human bondage. It is not the cultivation of *individual empathy*, but the affirmation of *collective will*, that can achieve the full development of human energy and creativity. [Italics added.][2]

The years since the toppling of the 'Gang of Four' in 1976 have brought a reappraisal of Maoist ideas, including views on empathy and development. A benchmark in this process was the motto, 'Seek truth from facts' (*shi-shi qiu shi*), popularized during the late 1970s. While Maoist truth was considered to be already discovered and only to require application through struggle, fact-based truth requires confronting alternative hypotheses as well as alternative applications. That is, struggle exists at both theoretical and practical levels. And to carry out the struggle successfully, one requires a flow of information concerning alternative points of view.

Telecommunications plays a vital role in providing this flow.

THE ELECTRONIC COURT

This vital role for telecommunications is based upon the collapse of the 'information float', as described in Naisbitt's *Megatrends*.[3] Electronic wizardry has virtually eliminated the 'float', or transmission lag, in the communication of information. This has two major implications. First of all, in the developed countries information such as news has been transformed from something which is sought after to something which is almost inescapable. Knowledge, the framework we use to make information useful, has not developed nearly as quickly. The user of information is squeezed between pressure to have complete, up-to-date information and pressure to devise means to channel it to avoid drowning in it. As the supply of information increases, the demand for knowledge intensifies.

The second implication is that it is now possible for experience, creativity and information resources to interact with unprecedented intensity. In the days of Peter the Great of Russia or China's renowned emperor, Ming Huang of the Tang Dynasty, an enlightened ruler assembled talented individuals in court. While in court they exchanged ideas and gave counsel on important matters. Use of their talents was often interspersed with long periods of leisure which left utilization of their talents at a low level. It was generally regionally rather than globally focused.

Electronic transmission has made it possible to pool talent with a rapidity and selectivity unimaginable in the days of Peter the Great, or even of Woodrow Wilson. It is possible to assemble talent globally on short notice. The selection of talent can be tailored to the problem at hand. Transmission lags are gone. Thus, superior talent can be assembled more rapidly, used more intensively and redeployed with less effort than ever before.

This potential for new intensity of contact implies the potential for creation of new knowledge which, in turn, may enhance the ability to benefit from future information.

As information becomes more plentiful and complete, it demands more from those who would organize it and put it to use. By their nature these heightened demands narrow the pool of talent competent to meet them. As demands escalate, national pools of talent capable of meeting them shrink. Innovative talent for the 'brain-intensive' side of information management is at a premium.

It must be cultivated and utilized, however, in ways which fit both national needs and the global scheme of things. It must *drive* development without *disrupting* it.

ROOTS AND BRANCHES

China's pool of people with talent for brain-intensive work could be viewed as having roots deeply set in their own cultural soil and branches reaching outside China. Both roots and branches provide the means to grow and develop. Each does so, however, in its own distinctive way.

The Cultural Revolution of 1966–76 disrupted the flow of knowledge by cutting China off from most of what was happening in the outside world. It also cut young Chinese off from much of their own past. To a greater or lesser extent, China's information environment from the 1950s through the 1970s was heavily laden with Maoist value-oriented communication. The Cultural Revolution represents an extreme. As Chu[4] points out, however, the presence of a constant stream of normative communication indicates that the translation of Maoist ideals into behavior falls far short of perfection. Such symptoms imply that multiple points of view, while submerged, were nonetheless present. They could not, however, undergo the kind of development possible through open, rigorous discussion.

After power changed hands in 1976, there was a period described with an expression freely translated as 'lingering fear' (*xin you yu ji*), when memories of the Cultural Revolution were fresh and its re-emergence seemed possible. This was followed by a period, beginning around 1980, when experiments with economic and political liberalization took place, accompanied by debates concerning what to keep, what to borrow or create and how to move forward without sacrificing national cohesiveness.

China's pool of talent works within this context of cohesive movement toward higher development. Information is the stimulus for this development. Telecommunication technology can transmit this stimulus throughout society. Therein lies its power to energize – and to disrupt.

While telecommunications technology has virtually closed the transmission gap, socio-cultural gaps remain wide open.

People who live within a socio-cultural system learn to cope with stimuli present in the system, even though their personalities and their coping strategies may vary widely. Edward Hall[5] has argued persuasively that to learn a culture intimately entails internalizing a system of rules as children and being unaware by the time we reach adulthood that our culture's rules are not everyone's rules. Culture shock occurs when we come in contact with other cultures and 'collide' with their systems of rules.

When cross-cultural contact is sustained or inescapable, two significant effects occur. The first is some degree of what Festinger termed cognitive dissonance.[6] Confronted with tension between new cognitive input and established cognitive material or beliefs, the individual attempts to reduce stress by rejecting or distorting new input or, otherwise, reconfiguring existing beliefs.

Difference in age, experience and personality may cause uneven levels of dissonance among members of the same culture. The young, for example, may feel less dissonance than their elders and deal with it more easily. Differing levels of 'computer anxiety' provide one such example. A more dramatic example might be the viewing of an American movie in which young people defy their elders, creating a 'dissonance gap' which is even more pronounced.

When suffered repeatedly at the group level, both cultural shock and unevenness in dissonance have the potential to cause an established socio-cultural system, whose purpose is to harmonize human diversity, to crack at the seams. Used to its full power, modern telecommunications can deliver unlimited amounts of unsettling stimuli.

The other effect, less immediate but equally salient, is adaptive reorientation. If reorientation is uneven, the resulting diversity in attitudes serves as a catalyst for constructive debate.[7] Too much unevenness, however, can be destructive. If, for example, a culture of lavish consumption like that portrayed on the television show *Dynasty* captures the imagination of a large number of people, their consequent discontent or squandering of resources might well lead to conflict in a society only recently moving away from the Spartan ideal.

China's pool of 'brain-intensive' talent has a dual task here. The first part of the task is the role of gatekeeper and filter for information to the general public. The general public, of course, receives a variety of outside information directly via telecommunications. Shortwave broadcasts are available throughout China. Consequently, the role of gatekeeper is not one of suppression or selection as much as one of organizing and interpreting information transmitted through telecommunications media. This process should 'seek truth from facts', but it should seek relevant truths among the stream of raw facts. This will enable facts to stimulate development of general knowledge and to foster the social cohesiveness necessary to use divergent viewpoints creatively.

The role of interpreter extends to the other part of the task, that involving specialized knowledge. The specialist in natural sciences, administration or marketing research needs state-of-the-art knowledge, delivered raw. Interpretation of the knowledge itself is up to the specialist. Anticipation of demand, prioritization of needs and obtaining the information all require study and interpretation of environmental trends. Here the specialist needs support.

This dual task will require a telecommunications system capable of intensive, flexible communication attuned to OECD standards for the specialized area of information. Information of general interest and utility will require less sophistication and less capital-intensity.

Tending roots for the time being will require less care and investment than growing branches.

CHANGING VIEWS OF THE DEVELOPMENTAL PROCESS

Since a principal focus of this chapter is development, a discussion of the process and measurement of development seems appropriate.

The concept of development is identified in a general sense with man's increased control over his natural environment. Anthropologists of the ecological-evolutionary school evaluate the degree of control with a measure called the techno-environmental advantage, the ratio of calories produced per calorie expended in production.[8] Forms of wealth allow humans to extend this advantage over time. Control of energy and, by extension, wealth, however, measures the availability of economic power but not its effects. Maslow's hierarchy of needs[9] provides a framework for evaluating a culture in terms of how well those living within it can satisfy their range of basic needs. Thus the dimension of well-being or quality of life is added to the dimension of economic power. Information has emerged as a resource which provides access, direct and indirect, to greater economic power and greater well-being. Its importance seems to be increasing.

This simplified sketch of development in terms of resource control and beneficial effects leaves out a key aspect: How do the resources achieve those effects and how can those effects be enhanced?

The 'how', that is the process of development, has been approached using three basic types of model.[10] Each type represents a compromise between parsimony and clarity on the one hand and faithfulness to complex realities on the other. The first two types of model, *mechanical* and *organic*, may be termed *universal*. The third type, *cybernetic*, may be termed *particular*.

Mechanical models view development in terms of universal patterns of linear causation. Economic factors play the key role, and development is measurable using indices such as per capita income, GNP, etc. Walt Rostow's analysis of economic development into five universal stages[11] is one example of this model. This mechanical approach ignores non-economic goals and cultural idiosyncrasies.

The *organic* approach also views development in terms of universal patterns of linear causation. It differentiates itself from the mechanical approach by stressing socio-psychological factors. In contrast to Rostow's vision of economies as gathering 'speed' until they 'take off' under their own power, the organic approach views development as movement up the hierarchy of complexity through change in individuals. Changes in individual cells eventually cause the entire organism to change.

Intellectually satisfying as they may be, such models have failed to satisfy those who attempt to use them as a guide for development.[12] Two likely reasons for dissatisfaction are as follows. First, these models are finely tuned to a limited set of factors (e.g. economic or psychological) while neglectful of other factors or interactions which are likely to affect development. Secondly, the view of a single, universal path toward a single developmental end may be viewed by the developing countries as a bid for promoting dependency upon the developed countries. It may also encourage attempts to transfer experience with inadequate adaptation to local circumstances.

The third approach, labelled *cybernetic*, counteracts these biases by allowing for more than one set of causal factors (multilinear causality) moving toward more than one possible end (probabilistic multifinality). It views society as a 'complex adaptive information-bound system'.[13] The process of development requires sustained, in-depth monitoring and adaptation, since inputs are complex and the path is not predetermined but negotiated through adaptation. What such a model loses in clear-cut precision it may gain in sensitivity to the process it represents and flexibility with respect to optional outcomes. The monitoring entailed in guiding development with such a model has implications for the use of telecommunications in information management and processing.

Throughout its period of organizational development (the 1920s through the 1950s) the Chinese Communist Party built and controlled its organization with communication which, though centrally directed, was carried out face-to-face in small groups. This type of communication and control is still useful for maintaining social cohesion, but the need to coordinate social, economic and technological development requires more complex feedback, much of which is not obtainable face to face in a small group setting. Higher levels of aggregation, more interactive reconfiguring of data, and tailored data dissemination on demand are not feasible in an organization built upon the cell-group model. They are, however, necessary to control many aspects of development. Telecommunications, coupled with data processing capability, can bring optimum pools of data and talent together for monitoring, diagnosis and adjustment at many levels. Restructuring of the governing process in China has begun to reflect this.

DISTINCT BUT OFTEN PARALLEL PATHS

While internal dynamics will absorb much of China's attention as development progresses, developments abroad, particularly among the OECD nations,[14] also merit a share of attention. One reason for this is that the OECD nations have the best-developed markets for brain-intensive labor, one of China's potential strong suits. Another is that concerns, problems and opportunities found in OECD information markets and telecommunications development may find parallels in China as she travels her own path to development. These parallels will be even more pronounced if China adopts the standards of OECD nations where possible in order to enhance her participation as buyer and seller in global markets. Some issues of concern among OECD nations are discussed below.

One issue has to do with the trade-offs between economies of scale and the advantages of diversity. Standardized telecommunication services can be run as regulated monopolies to maximize economies of scale. This minimizes the cost burden, making these services accessible to the widest number of customers and freeing funds to be diverted into other economic activities. This, however, minimizes funds available for research and development. Diversity, on the other hand, tends to breed competition and innovation, though at a higher short-term cost to the customer.

H. Ergas of the OECD's Advisory Unit on Multidisciplinary Issues, wrote the following in 1987:

> There can be little doubt that engineering cost estimates do demonstrate substantial economies of scale and scope to the telecommunications network. This conclusion is . . . not substantially altered by the availability of new technologies.
>
> It is, as yet, too early to analyse the effects of competition on the provision of basic services. Nonetheless, the experience to date suggests that the transition to competition creates a host of problems [which] arise mainly from two sources: the asymmetry in market power between the established supplier and new entrants; and the effects of prior investment and financing decisions on the established supplier's cost structures.[15]

While Ergas sees the introduction of competition into natural monopolies (defined by economies of scale) in basic telecommunications services in inconclusive terms, he does not view the natural monopoly as extending beyond basic services:

> There is not evidence of natural monopoly in VAN or CPE markets; the case for restricting competition in these markets is therefore highly problematic.[16]

The level and extent of competition to be introduced into China's growing telecommunications equipment and services industry[17] may be governed by such concerns as whether a natural monopoly exists and whether services are basic or value-added in addition to political considerations. Even if such factors do not weigh heavily internally, they will determine the patterns of competition externally.

Another issue is the difficulty users have in accessing the full range of required services, particularly transmission of information in voice, data and image forms from different networks. The ISDN (Integrated Services Digital Network) concept is an attempt to eliminate costly interfacing and make these forms of information equally accessible through an integrated system.[18] Less developed systems, like China's could integrate earlier at less cost.

Another area of concern is the social cost/benefit of improved communication technologies. Frederick Williams wrote in 1982:

Despite the glitter, promise, or threats of the new communications technologies, nothing is so important as is their consequences upon our daily lives. These consequences do not tend to be entirely new activities for us, but appear more as changes in traditional services and the institutions which supply them.[19]

Communications technologies have the potential to effect profound changes in the life and institutions of China, a country which the Revolutionary statesman, Sun Yat-sen, once characterized as being uncohesive as 'a tray of loose sand' (*yi pan san sha*). China is no longer a tray of loose sand, but her fully modernized form has yet to emerge. Her policy decisions will determine this form.

CONCLUSIONS AND RECOMMENDATIONS

Having emerged from the isolationism of the Cultural Revolution (1966–76), China needs to seek a path to development through participation in global information markets. Such participation must be planned so as to minimize information dependency.

China's pool of 'brain-intensive' talent will play a critical role in this process both domestically and externally. Domestically it will serve to filter and adapt input, while externally it will serve as the link to China's participation in global information markets. Telecommunications technology allows this pool of talent to be used with unprecedented intensity and efficiency.

Of the approaches available, the *cybernetic* type of model seems best suited to China's developmental effort because of its responsiveness to the full range of relevant factors. The monitoring required to use this type of model will require creation of extensive and sophisticated linkages between telecommunication and data processing networks.

Thus, monitoring China's development will also catalyze it.

Notes

1. See Daniel Lerner, *The Passing of Traditional Society: Modernizing the Middle East* (Glencoe, IL: Free Press, 1958). Also see Lerner 'Technology, communication and change', in Daniel Lerner and Wilber Schramm (eds), *Communication and Change: The Past Ten Years and the Next* (Honolulu: University Press of Hawaii, 1976).
2. Godwin Chu, *Radical Change through Communication in Mao's China* (Honolulu: University of Hawaii Press, 1977) p. 254.
3. See John Naisbitt, *Megatrends: Ten New Directions Transforming Our Lives* (New York: Warner Books, 1982) pp. 11–38.
4. Godwin Chu, op. cit., p. 258.
5. See Edward T. Hall, *Beyond Culture* (New York: Anchor Press, 1976).
6. See Leon Festinger, *Theory of Cognitive Dissonance* (New York: Harper & Row, 1957).
7. For a detailed analysis of the development of this process in China, see Andrew Nathan, *Chinese Democracy* (New York: Alfred A. Knopf, 1985).
8. See Marvin Harris, *Culture, Man and Nature* (New York: T. Y. Crowell, 1971) pp. 203–217.
9. A. H. Maslow, *Motivation and Personality* (New York: Harper & Row, 1954). Maslow categorizes human needs using a five-level hierarchy with physiological needs at the bottom and then, in ascending order, needs for safety, affiliation, esteem and self-actualization.
10. Here the categorization used by Majid Teheranian in an editorial entitled 'Communications and development: the changing paradigms', in *Communications and Development Review*, (Vol. 1, Nos 2 and 3, Summer–Autumn 1977) has been followed.
11. See W. W. Rostow, *The Economic Stages of Growth* (2nd edn, Cambridge University Press, 1971). Rostow's Stages are: (1) the traditional Society, (2) the preconditions for Take-off, (3) the Take-off, (4) the drive to Maturity, and (5) the age of High Mass Consumption.
12. The editorial by Teheranian, cited above, was motivated by this dissatisfaction.
13. Teheranian, op. cit., p. 3.
14. The Organization for Economic Cooperation and Development, whose membership includes most of the industrially advanced world.
15. H. Ergas, 'Regulation, monopoly and competition in the telecommunications infrastructure', in *Trends of Change in Telecommunications Policy* (ICCP (Information Computer Communications Policy)) publication No. 13: Paris: OECD, 1987), pp. 46–7.
16. Ergas, op. cit., p. 47.
17. For a detailed overview of China's telecommunications industry see the seventeen-part series entitled 'China's industry profile: telecommunications' appearing in *Business International: Business China*, 7 November

1985–28 July 1986.
18. This concept is discussed in A. Hutcheson Reid, 'The Integrated Services Digital Network: a presentation of policy issues', in *Trends of Change in Telecommunications Policy*, (ICCP (Information Computer Communications Policy)) publication No. 13: Paris: OECD, 1987.
19. Frederick Williams, *The Communications Revolution* (Beverly Hills: Sage Publications, 1982) p. 12.

Section V
Economic Policy

11 United States Bilateral Foreign Aid and Multilateral Aid: A Comparison

David Porter

The empirical literature has adopted two approaches, based on distinctly different motives, to explain the allocation of United States bilateral foreign aid. The first approach stresses the interests and foreign policy goals of the United States, and operationalizes a number of models to capture different conceptualizations of United States national interest. In contrast the second approach stresses altruistic motives by operationalizing a model designed to capture the basic human and economic needs of the recipient state.[1]

With some reservation the literature generally supports the conclusion that United States foreign aid policy reflects national self-interests, rather than the interests of the recipient state.[2] A question that has not been empirically tested is whether the allocation of multilateral aid through international governmental organizations (IGOs) also reflects the national interest of the United States. The hypothesis of interest is whether the level of United States donations to the following multilateral aid agencies, the UNDP, UNFPA, UNICEF, WFP, IDA, IFAD, IFC and the IBRD, has permitted the United States to 'capture' or dominate the decision-making process and control the allocation of multilateral aid. In effect this hypothesis suggest that the United States uses multilateral aid allocations to support or reflect the national interests of the United States.

RESEARCH DESIGN

The empirical foreign aid research adopts a uniform research design which, with some variations, is adopted here. The research design is based on the theoretical constructs of the rational choice paradigm;

a series of decision-making models are specified and then tested statistically for validity. To meet these basic requirements it is necessary to specify the decision-maker of interest, identify the dependent variables, and operationalize a set of independent variables that capture the relative utility of each decision-making strategy being tested (Holt and Richardson, 1976).

The decision-making units of interest are the United States and the multilateral agencies noted. Each decision-making unit was considered separately, and the results compared to determine the preferences and priorities of each. The analysis of United States allocations has been limited to United States economic foreign aid administered by the Agency for International Development (USAID).

The foreign aid literature tends to operationalize total aid allocations as the dependent variable. But a portion of the literature has tested whether allocation patterns vary across foreign aid programs. Kato's research in particular suggest that there is a significant variance in the allocation of military and economic aid.[3] Since multilateral aid is economic in nature it would be inappropriate to compare total United States allocations (military + economic) with multilateral allocations.

USAID allocations are operationalized as total allocations and by budget category. USAID's annual budget is divided into five categories of economic assistance: (1) population planning, (2) health, (3) education and human resources, (4) selected development projects, and (5) agriculture, rural development and nutrition. Each category effectively represents a functional division of labor or a different foreign aid mission. The hypothesis of interest is whether there is a functional relationship between the allocation of bilateral economic aid by budget category and multilateral allocations by IGO's with similar functional responsibilities.

The data set consist of those Latin American states receiving bilateral aid from the United States over a two year period – 1982 and 1983. Latin America was selected because it is usually assumed that the United States has special interests in the region and is more active in pursuing those interests. Consequently, it is reasonable to assume that the United States is more likely to control multilateral allocations for this region than for Africa or Asia.

There are three potential methodological difficulties that must be addressed. First is the potential for simultaneous causation resulting from a relationship between the dependent variables that measure the level of foreign aid, and the independent variable the GDP of the recipient state. To avoid the potential for simultaneous causation, the level of foreign aid is removed from the independent variable GDP. (adjusted GDP = GDP − aid).

However, the model operationalized to measure the economic self-interest of the donor state includes three measures related to GDP: (1) mining as a percentage of GDP, (2) manufacturing as a percentage of GDP, and (3) agriculture as a percentage of GDP. Since it is not possible to determine the relationship between the level of aid and a specific economic activity, no corrective action can be taken. Consequently, the potential for a simultaneous causal relationship must be acknowledged. But, it seems unlikely that any such relationship will be strong enough to cause serious results (Pindick and Rubinfeld, 1981, pp. 152–161 and 191–199).

The second potential methodological problem stems from the variance in the size of recipient states. It is quite possible that the error terms associated with large recipient states such as Mexico will have larger variances than the error terms of smaller recipient states, such as Jamaica, causing a heteroscedastic relationship.

To test for heteroscedasticity, the error terms were converted into absolute values and regressed against the independent variables. A random sample was tested and the results were negative. Consequently, it is assumed that variance of the error terms is consistent across cases, and no corrective action was taken (Pindick and Rubinfeld, 1981, pp. 140–52)

The third methodological problem concerns the measurement of the dependent variables. The question is whether to operationalize the level of aid as a gross measure, or as a per capita measure to reflect the relative size of the recipient state by adopting per capita measures of foreign aid. The objective of the transformation is to adjust the dependent variable to reflect each recipient state's individual 'relative need' for economic aid. While it is appropriate to measure the relative need of the recipient states the comparative research literature raises serious questions concerning the reliability and consistency of per capita measures when using ordinary least squares. Fortunately, most of the concern is limited to those cases where per capita measures are operationalized as both dependent and independent variables. Only one independent variable is operationalized as a per capita measure, per capita GDP, and corrective action has already been described.

Eric Uslaner, in his article 'The Pitfalls of Per Capita' develops two decision rules to guide the researcher in the use of per capita measures: (1) where the interest of the researcher 'involves explicit comparisons among cases . . . such as relative deprivation', and (2) where one has 'prior knowledge that the relative independent variables actually employed by decision-makers . . . includes standardized measures', such as per capita transformations (Uslaner, 1976, p. 132). To explain the allocation of foreign aid, it is necessary to measure and compare the relative level of United States national interest and the relative need of each recipient state. In effect this provides for a comparison between cases, which is precisely the type of research situation referred to in Uslaner's first decision rule.

In reference to Uslaner's second decision rule, it is appropriate to assume that foreign aid decision-makers base some portion of their foreign aid allocation decisions on per capita measures of the recipient state. Per capita measures are commonly reported and used by both domestic and multilateral aid agencies as a means of classifying and determining the relative need of recipient states. For these reasons the per capita measure is adopted.

THE MODELS AND INDICATORS

The models and indicators adopted are derived from the research of McKinlay and Little. McKinlay and Little found no support for the recipient need model.[4] Paul Mosley has questioned McKinlay and Little's findings, noting errors in the recipient need model specification. Using a respecified model and two-staged least squared statistical analysis Mosley was unable to reject or accept the recipient need model (Mosley, 1981, p. 252). However, Mosley's respecified model failed to operationalize several measures of basic human needs suggested by Hicks and Streeten.[5] The recipient need model operationalized here attempts to expand Mosley's model by adding the indicators suggested by Hicks and Streeten.

A similar approach was adopted concerning McKinlay and Little's political ideology and stability model. McKinlay and Little combined the two phenomena into one model. However, according to Geller democracy and stability are two independent phenomenon and combining the two into one model may cause spurious results (Geller, 1982). To avoid spurious results Bolland's measures of democratic development, which have been found to be both reliable and consistent, were adopted.

The dependent variables include United States economic foreign aid administered by USAID, total and by buget category. The level of allocations made by the UNDP, UNFPA, UNICEF, WFP, IDA, IFAD, IFC and the IBRD is operationalized to capture multilateral allocations. All dependent variables are per capita measures and are in United States dollars. The multilateral agencies and USAID budget categories being tested include:

1. USAID total economic foreign aid – all forms of military aid removed.
2. USAID by budget category – Agriculture, Rural Development and Nutrition.
3. USAID by budget category – Population Planning.
4. USAID by budget category – Health.
5. USAID by budget category – Education and Human Resources.
6. USAID by budget category – Selected Development Activities.
7. Food for Peace Program – PL480.
8. UNDP – UN Development Program.
9. UNFPA – UN Fund for Population Administration.
10. UNICEF – UN Children Fund.
11. WFP – World Food Program.
12. IDA – International Development Association.
13. IFAD – International Fund for Agricultural Development.
14. IFC – International Finance Corporation.
15. IBRD – World Bank (IBRD) disbursements.

To test the hypothesis that the distribution of economic foreign aid is proportional to the humanitarian and basic human needs of the recipient state the recipient need model is operationalized to measure the economic and basic human condition need of the recipient state:

1. The balance of payments of the recipient state.
2. The gross population of the recipient state.
3. Calorie intake as a percentage of total daily requirements.
4. Infant mortality rate for first year.
5. Life expectancy at birth.
6. Per capita gross national product.
7. Population per physician.
8. Food Index.
9. World Bank (IBRD) economic development classification.

To test the hypothesis that the distribution of economic foreign aid is proportional to the security interest of the United States, the security interest model measures the bilateral security relationship between the recipient state and the United States. One indicator commonly used to measure bilateral security relationships – defense and other treaties – has been omitted due to classification problems resulting from the Rio Treaty and the OAS, which include all members of the database with the exception of Belize:

1. The gross level of the United States military assistance.
2. The presence of United States troops, military technicians, or advisers operationalized as a dummy variable.
3. The presence of United States bases operationalized as a dummy variable.
4. The United States arms transfers including subsidized and non-subsidized arms sales operationalized as a dummy variable.

There was some difficulty finding reliable data concerning military advisers, technicians and the location of bases; for this reason these indicators were operationalized as dummy variables and there is concern that dummy variables may not be sensitive enough to capture the bilateral security relationship between the United States and the recipient state. To address this concern a second security model is operationalized that replaces the indicators for United States arm transfers, bases and personnel with a security index. The presence of each variable is assigned the value '1', the absence the value '0'. The values are then added and divided by '3', the number of ties being measured. The index created provides a measure ranging from 1 to 0 – the closer the indexed score is to 1 the tighter the security relationship between the United States and the recipient state. Both versions of the security interest model will be tested against the dependent variables.

To test the hypothesis that the distribution of economic foreign aid is proportional to the potential geopolitical power of the recipient state, the geopolitical model is designed to measure the power capabilities or potential of the recipient state:

1. The gross population of the recipient state.
2. Adjusted domestic product of the recipient state.
3. Recipient state's gross international reserves.
4. Recipient state's gross military expenditures.
5. Recipient state's military expenditures as a percentage of total GNP.
6. The size of the recipient state's military establishment in thousands

of members.
7. The size of the recipient state's military establishment as the percentage of total population in the military.

To test the hypothesis that the distribution of economic foreign aid is proportional to the economic self-interest of the United States, the economic self-interest model measures the economic relationship between the United States and the recipient stage, and the potential for United States interest in the recipient state's economy:

1. Exports of fuels, minerals, and metals from the recipient state to the United States as a percentage of total exports.
2. Exports of other commodities from the recipient state to the United States as a percentage of total exports.
3. Domestic investments.
4. Domestic savings.
5. Percentage of GNP in manufacturing.
6. Percentage of GNP in industry.
7. Percentage of GNP in services.

To test the hypothesis that the distribution of economic foreign aid is proportional to the ideological consistency between the United States and the recipient state, the political ideology model is operationalized with two indicators designed to replicate Bollen's measures of popular sovereignty and political liberties (Bollen, 1980):

1. Political rights index, with a score of 1 indicating a high degree of political rights and a score of 7 indicating a low level of political rights.
2. Civil liberties index, with a score of 1 indicating a high level of civil liberties, and a score of 7 indicating a low level of civil rights.

To test the hypothesis that the distribution of economic foreign aid is proportional to the bilateral relations between the recipient state and the USSR, the containment of communism model is designed to capture the bilateral relationship between the USSR and the recipient state:

1. The presence or absence of USSR arm transfers.
2. The presence or absence of USSR defense treaty.
3. The presence or absence of USSR treaties of friendship.
4. The presence or absence of a domestic communist party.
5. The presence or absence of USSR military bases or base access.

6. The presence or absence of USSR technicians or military advisers.
7. The presence or absence of Cuban technicians or military advisers.

A second containment model is operationalized that replaces the indicators for USSR arm transfers, defense treaties, treaties of friendship, military bases, technicians or military advisors, domestic communist party and Cuban technicians or military advisors with an indexed measure of the bilateral relationship between the USSR and the recipient state. The presence of each variable is assigned the value '1', the absence the value '0'. The values are then added and divided by '7', the number of ties being measured. The index created provides a measure ranging from 1 to 0 – the closer the indexed score is to 1 the closer the relationship between the USSR and the recipient state. Both versions of the containment model will be tested against the dependent variables.

EMPIRICAL RESULTS

The primary hypothesis tested was whether the United States has captured the multilateral allocation process. The hypothesis will be supported if one of two conditions is met. First, if there is a relationship between total United States bilateral allocations and the distribution of aid by multilateral sources. To test for this condition it is necessary to determine the level of correlation between USAID allocations and multilateral allocations. The second condition is whether there is a relationship between multilateral aid allocations and the models operationalized to capture the United States national interests and needs of the recipient state. To test for this condition it is necessary to regress multilateral and USAID allocations against the six models operationalized. Each model and dependent variable will be considered independently and the results compared.

The first condition was tested by comparing the allocation pattern of USAID foreign aid with the allocation pattern of multilateral foreign aid for the eight IGOs identified. A very strong relationship was found between bilateral and multilateral aid allocations for the IBRD, and a moderate relationship was found for UNDP. The results for the other multilateral agencies tested did not find any additional reliable and consistent relationships.

The captured hypothesis is rejected for the IFAD, IDA, WFP and

UNICEF. For these multilateral agencies it does not appear that the United States greatly influences the multilateral foreign aid allocation decision-making process. The hypothesis is accepted for the IBRD and UNDP. The allocation pattern for these two agencies showed a positive relationship with USAID allocations indicating that the United States has a significant level of influence over the allocation of funds. However, it must be remembered that these results are limited to Latin America and could change when other regions are added to the analysis.

With the exception of the USAID Health budget category for 1983 a strong and significant relationship was found between all USAID allocation categories and IBRD allocations, for both years. The USAID Health category was found to be very significant for 1982 but was not significant for 1983. US budget allocations under the Health category decreased significantly between 1982 and 1983. This may explain the strong relationship reported for 1982 and the lack of a relationship in 1983.

The relative strength of the relationship between IBRD and USAID allocations, as measured by R squared, varied from a high of .78 for USAID Health allocations in 1982 to a low of .40 for the USAID Population budget category. The value of the unstandardized regression coefficient (beta) ranged from a low of .53 for USAID Education in 1983, to a high of .69 for USAID Health in 1982. The only significant change between 1982 and 1983 was in the USAID budget category for Health programs.

The limited range of beta is considered to be an indication that the level of United States influence over IBRD allocation decisions is fairly consistent across budget categories and developmental strategies. Given the often reported influence and virtual veto over IBRD decisions exercised by the United States these results are not considered surprising.

A positive relationship was found between the UNDP and total USAID allocations. The recorded relationship was moderate, but the results still indicate that the United States significantly influences UNDP allocation decisions. For total USAID allocations the relationship was significant for both years with the value of R squared increasing from .38 in 1982 to .49 in 1983.

However, upon further investigation it appears that a significant portion of the total USAID variable's explanatory power resulted from a close relationship between UNDP allocations and funds allocated under the USAID budget category for Population programs.

For both years the values of R squared and beta are greater for the USAID Population variable, than for the total USAID allocations, indicating a stronger and more responsive relationship.

The second condition will be met if multilateral allocations reflect the interests of the United States rather than the recipient state. The results for total USAID allocations and by budget category tend to confirm earlier research in their support for the donor national interest explanation. There were, however, some interesting variations between this research and earlier research in the field. Still, even given the variation between this research and previous research in the field, the results are clear, and the hypothesis that USAID allocations are made on the basis of the needs of the recipient state is rejected and the donor interest hypothesis is accepted.

The results for the recipient interest model were considerably different for the multilateral agencies. The model was found significant for at least one year for every agency except for the IBRD and the IFAD. But the direction of the relationship does not support the recipient need hypothesis. The value of beta for life expectancy was $-.69$ for 1982 and $-.80$ for 1983 for UNICEF allocations. The value of R squared was .56 and .71 respectively. For UNFPA the value of beta is $-.47$ for 1983 and R squared was .22. The relationship was not significant for 1982. What must be noted about these results is that the value of beta is negative which means that the greater the value of the variable, or the longer the life expectancy, the higher the level of aid allocated to that recipient state.

To accept the recipient need hypothesis it is necessary that the results be significant and the relationsnip between life expectancy and the allocation of aid be positive. UNICEF and UNFPA fail to meet the second portion of this test and the hypothesis is rejected for these agencies. The hypothesis is accepted for the WFP, however. For the WFP the value of beta was a positive .58 for both years. The explanatory power of the life expectancy indicator was moderate with a R squared value of .34 for 1982 and .31 for 1983.

The recipient need hypothesis is also accepted for the IDA because of a positive relationship between infant mortality rates and the allocations of foreign aid. The value of beta was .73 for 1982 and .65 for 1983, the value of R squared .53 and .42 respectively. The value of these measures indicates a strong and responsive relationship between infant mortality rates and the allocation of aid by IDA. Based on the results of this study it appears that the WFP and IDA are responsive to the level of infant mortality in the recipient states.

Four multilateral agencies, UNDP, UNFPA, UNICIEF and the IFC, showed a relationship between population size and the allocation of foreign aid. For the UNDP the value of beta was a positive .48 for 1982, indicating that the larger the size of a recipient state's population the more likely the state is to receive funds through the UNDP. For the UNFPA, UNICEF and the IFC the value of beta was negative, indicating that these agencies tend to allocate funds to the smaller or less populated recipient states. These results may be an indication that there is some level of specialization based on the relative size of the recipient state across the multilateral agencies. This is speculation, however, and there is no evidence from this study to support this proposition.

While the recipient need hypothesis is accepted for only two multilateral agencies, a significant variance was noted in the empirical findings across bilatateral and multilateral allocations of foreign aid. This is interpreted as an indication of independence on the part of the multilateral agencies. There is no evidence, based on the recipient interest model, that the United States has captured or dominates the allocation of multilateral aid. However, it should be noted that the model was not found significant for the IBRD, which is the IGO most likely to be influenced by the United States.

The security interest model was significant for both years with the index of United States security interests the most salient indicator. For 1982 the security interest model was significant for all but the Health budget categories. The level of explanatory power varied from a high of .47 for USAID Agriculture, to a low of .28 for USAID Education. R squared for total USAID allocations was .40.

These figures changed significantly for 1983; R squared for total bilateral allocations dropped to .20 and the beta measure dropped from .63 to .45, indicating a decrease in the explanatory power of the model and in the responsiveness of the dependent variable to changes in the independent variable. With the exception of the USAID Population category, which remained constant, the other categories of United States bilateral aid followed the same pattern, a sharp reduction in explanatory power and sensitivity between 1982 and 1983.

Given the public policy statements of the Reagan administration, this pattern was not anticipated. Rather it was anticipated that the explanatory power of the security model would increase as the United States foreign policy focused on Central America. However, it must be remembered that only United States foreign aid allocated through

USAID is being considered. It is quite possible, and there is some evidence to suggest, that aid to Central America has increased significantly outside of USAID channels. For example, increases in military aid or special financial concessions would not be captured by the aid indicators operationalized by this study. Still, the security interest model was found to be the most powerful model in explaining the allocation of USAID funds across the Latin American region.

In general, the economic interest model was not capable of explaining multilateral aid allocations for 1982, with the exception of a relationship between the level of manufacturing and the allocations of UNDP funds. The model was more successful for 1983, and a strong relationship was found between the allocations of aid by UNFPA, UNICEF, IDA, IFAD and the IBRD.

The last foreign aid model to be considered is the containment of communism model. The model was not significant for any dependent variable, bilateral or multilateral, for either year. And the hypothesis that aid is allocated to contain communism in Latin America is rejected. At first these results may be confusing given the United States' focus on stopping the spread of communism in Central America. But it must be remembered that the model assumes that the recipient states in question are communist or are receiving foreign aid from the communist bloc. For Central America neither case is true, and the effects of the increased aid to Central American states is captured by the security interest model.

A secondary hypothesis that was also of interest was whether there was a functional relationship betwee USAID programs, as measured by USAID budget categories, and multilateral agencies with similar foreign aid missions, such as the UNFPA and the USAID budget category for Population planning. It was thought that there might be some cooperation across bilateral and multilateral programs. However, the empirical results were negative and the hypothesis that there is a functional relationship between multilateral allocations and USAID budget categories is rejected.

Several of the USAID budget categories were found to be correlated with multilateral allocations, but the relationships were not reflective of any clear functional division of labor. For example, the level of the recipient state's domestic investments was found significant for five of the eight multilateral agencies for the year 1983; the indicator was not significant for 1982. The explanatory power of the indicator, as measured by R squared, varied from a high of .49 for the IBRD to a low of .19 for IDA. The value of beta remained fairly constant across the multilateral agencies, ranging from a high of .67 for the IBRD to a low of .47 for UNFPA. Both these measures indicate a strong and sensitive relationship between the allocation of multilateral aid and the level of recipient state domestic investment, but do not indicate a reliable and consistent relationship based on functionalism.

The positive findings between the IBRD, the UNDP and USAID budget categories also illustrate the point. While several relationships were found there is no indication that these relationships are based on functionalism or represent any level of policy coordination. If the bilateral and multilateral allocations were coordinated or controlled one would expect a strong relationship between one or two budget categories and the multilateral allocations for IGOs with a similar mission. For the IBRD a strong relationship was found for all budget categories, except for Health, for 1983. And while the Population budget category showed a very strong relationship with UNDP allocations, the primary mission of the UNDP is development not population control. Consequently the relationship does not support the hypothesis. There is no evidence to suggest a significant level of program or policy coordination between USAID and the multilateral agencies tested.

When combined with the distinctly different pattern of allocations reported for the multilateral agencies and for the economic self-interest and recipient interest models, these results are considered an important indication that the multilateral agencies have a relatively high level of decision-making independence.

CONCLUSIONS

For the bilateral aid models the results are consistent with McKinlay and Little in that the donor interest model was accepted, and no support was found for the recipient need model (McKinlay and Little, 1979, pp. 242–50). The most interesting variation, for bilateral allocations, is the saliency of the geopolitical model. The model was not significant for any dependent variable, bilateral or multilateral, for either year. In contrast, McKinlay and Little found the geopolitical model to be the most salient in explaining the allocation of bilateral United States foreign aid. The Central American crises may have caused these results. During the 1980s the United States increased aid to the Central American states, which explains some of the validity of the security interest model. But the Central American states have limited geopolitical power, and increased allocations to these states would tend to decrease the validity of the geopolitical model. Of more import is that both the geopolitical and security interest models reflect the interest of the donor state and not the needs of the recipient state.

The fact that this research is not an exact replication of McKinlay and Little's models increases the reliability of the findings. The variations in research design should have improved the accuracy and sensitivity of the recipient need model. Of particular import this research considered only economic aid and divided USAID allocations by budget category. This approach represents a best case research design. If the United States allocated aid based on the needs of the recipient state one would expect a relationship between the needs of the recipient need model and one or more of USAID's budget categories, such as Population or Health, but no relationship was found. Unless there is a major policy change, the empirical results are strong and consistent enough to assume that the allocation of United States foreign aid reflects United States national interest.

For the multilateral models the empirical results are mixed. It was hoped that the results would be clear enough either to reject or accept the hypothesis that the United States has captured the multilateral allocation process with confidence. However, the hypothesis was accepted for the IBRD and UNDP, but had to be rejected for the other multilateral agencies tested. The relative strength or the level of influence the United States has over the IBRD and UNDP cannot be determined with accuracy because of the inability of the six foreign aid decision-making models to identify the areas of cooperation between bilateral and multilateral allocations.

It must be assumed that the IBRD and the UNDP use some of the same decision-making criteria in the allocation of foreign aid as USAID. Otherwise there would not have been a relationship between bilateral allocations and IBRD and UNDP allocations. It appears clear that the indicators and models operationalized, which were common to the field, were not adequate to capture or describe fully the bilateral or multilateral decision-making process. Additional research is necessary to identify the full range of decision-making criteria utilized in the allocation of foreign aid.

Still, this study has found a significant, and perhaps surprising, level of independence between bilateral and multilateral foreign aid allocations. And while the level of independence cannot be determined, it appears clear that the United States does not control and has not captured the multilateral allocation process for all, or even most, of the multilateral agencies with decision-making authority over multilateral foreign aid allocations.

Notes

An earlier version of this paper was presented at the international conference of the Association for the Advancement of Policy, Research and Development in the Third World. I would like to thank Dr Mekki Mtewa for his helpful comments and Lisa Malmer for her assistance. This research was supported in part by a grant from Youngstown State University, though the opinions and findings reported do not necessarily reflect the views of the University.

1. Both the normative and the empirical literature provide discussions of the relative merit of the donor interest and recipient need approaches. For a normative discussion see Morgenthau (1962); for an empirical consideration see McKinlay and Little (1979) and Mosley (1980).
2. A number of researchers have empirically tested various forms of the donor interest and recipient need models, and while there are important differences in research design and objectives their conclusions tend to support the donor interest approach. The primary exception is Mosely and his results were inconclusive – see Kato (1969), Cohn & Wood (1980), Schoultz (1981), Mosley (1981), and McKinlay and Little (1977 and 1979).
3. Kato (1969) in his research considered total United States foreign aid and military and economic aid independently. While his results were consistent in that both forms of aid were allocated to reflect the interests of the donor state Kato noted several significant differences in the explanatory power of the independent variables, suggesting that the two forms of aid were utilized to achieve different policy objectives.
4. McKinlay and Little have published two papers, one in 1977, the second in 1979, testing the donor interest and the recipient need models. While there are differences between the two papers, particularly in reference to the specification of the dependent variable, the results have been consistent in their support for the donor interest model.
5. Mosley's (1981) recipient need model tends to stress economic needs over basic human needs. Hicks and Streeten (1979) stress the importance of meeting basic human needs in their discussions. The model operationalized for this project combines both sets of indicators to measure the needs of the recipient state.

References

BOLLEN, K. 'Issues in the comparative measurement of political democracy', *American Sociological Review* (June 1980), pp. 370–90.

COHN, S. and WOOD, R. 'Basic human needs programming: an analysis of Peace Corp data', *Development and Change*, Vol. 11 (1980), pp. 313–32.

GELLER, D. 'Economic modernization and political instability in Latin America: a causal analysis of bureaucratic authoritarianism', *Western Political Science Quarterly* (March 1982), pp. 33–49.

HICKS, N. and STREETEN, P. 'Indicators of development: the search for a basic needs yardstick', *World Development*, Vol. 7 (1979), pp. 567–80.

HOLT, R. and RICHARDSON, J. 'Competing paradigms in comparative politics'. In Holt, Robert and Turner, John, *Methodology Of Comparative Research*. New York, The Free Press, 1976. Chapter 2.

HUNTINGTON, S. 'Foreign aid for what and for whom', *Foreign Policy*, 1974.

KATO, M. 'A model of US foreign aid allocations: an application of a rational decision-making scheme'. In Mueller, John (ed.), *Approaches to Measurement in International Relations*. New York, Appleton-Century-Crofts, 1969, pp. 198–215.

LOWENTHAL, A. F. 'Foreign aid as a political instrument: the case of the Dominican Republic', *Public Policy*, No. 14 (1965), pp. 141–60.

McKINLAY, R. D. and LITTLE, R. 'A foreign policy model of US bilateral aid allocation'. *World Politics* (October 1977), pp. 58–86.

McKINLAY, R. D. and LITTLE, R. 'The US aid relationship: a test of the recipient need and the donor interest models', *Political Studies* (June 1979), pp. 236–50.

MORGENTHAL, H. V. 'A political theory of foreign aid', *American Political Science Review*, No. 56 (1962), pp. 301–9.

MOSLEY, P. 'Models of the aid allocation process: a comment on McKinlay and Little', *Political Studies* (June 1981), pp. 245–53.

PINDYCK, R. and RUBINFELD, D. *Econometric Models and Economic Forecasts*, 2nd edn. New York, McGraw-Hill, 1981.

SCHOULTZ, L. 'US foreign policy and human rights violations in Latin America', *Comparative Politics* (January 1981), pp. 149–70.

USLANER, E. 'The pitfalls of per capita', *American Journal of Political Science* (February 1967), pp. 125–33.

12 Economic Development in Latin America: The Brazilian Experience

Eufronio Carreño Román

A number of Latin American countries had a sufficient level of development by the 1980s to be referred to as the Newly Industrialized Countries. They reached that status by pursuing and implementing a series of economic policies that had begun in the 1930s. This paper will examine these policies and the experience of Brazil in its quest for development. The argument put forward is that the government played an active role and used nationalism as an ideology and rationale for its policies. The first part of the paper presents a review of the relevant literature. The second part presents the argument. The third part presents evidence that supports the argument, followed by the conclusions which may be drawn.

REVIEW OF THE RELEVANT LITERATURE

Rostow[1] presents a system of five stages of growth with more variables. He argues that changes of stages of growth are the result of the interaction of individuals in a society. This individual behavior 'is not uniquely determined by economic considerations', but culture, social fabric and political forces also affect the performance of the dynamic growth. Production is determined by individuals choosing the path that balances conflicting alternatives. However, the idea of stages was started as early as Marx.[2] In his materialistic interpretation of history, Marx perceived economic development as a dialectic process of changes in the mode of production. The relationship of production embodies the institutional framework under which economic development takes place. In the very long run, the development of productive forces breaks up the relations of production to move the development into another mode of production. Dependency theorists argue that underdevelopment and the problems

189

of progress for developing countries may be caused, over time, by their economies coming into contact with developed capitalistic systems.[3]

A difference should be made between economic development and economic growth. Economic development implies a qualitative change in the economic system, a change in the way production and consumption of goods and services takes place in an economy. New sectors are included in the economic activity which generates an increase in the rate of per capita income. Productivity also increases as more efficient techniques of production are introduced to the economy. This process is carried out within a specific institutional framework which changes at each stage to foster the development process; furthermore, this can come slowly in response to innate economic forces, or it can be consciously accelerated by appropriate government policies. Finally, it can be induced by exogenous shock.

Economic growth on the other hand, involves a quantitative rather than a qualitative change. It is the growth of per capita income. This does not necessarily require a change in the institutional framework; it can happen within the existing one, i.e. growth is a quantitative accumulation.

The growth models described previously assume that the government will not intervene in the economy and will maintain international economic links, thus individual producers and consumers, under a *laissez-faire* policy, will seek to maximize income. On the other hand, the Marxian and dependency models assume that government policies are aimed at justifying and/or maintaining the established economic status quo.

The proposition considered here is that in the latter part of the twentieth century in Latin America, it was the appropriate economic policies consciously implemented by governments that used nationalism as an ideological justification to alter the international economic links that brought about movements into higher stages of economic development.

THE ARGUMENT

The argument assumes that the country is endowed with natural resources and a positive rate of population growth. It is couched in a framework of stages of development,[4] beginning when the country in focus is in the early stage (ES) of development. The government has been following the appropriate economic policy, namely, *laissez-faire* policies along the lines of the international comparative advantage, which has allowed the economy to move to this stage. The modern sector exports some commodities using modern technology copied from the more advanced economies, while the traditional sector remains using rudimentary traditional techniques of production. Thus international trade gives rise to a monetary market-economy, which coexists alongside the traditional (non-monetary) economy. However, the movement to the next stage of development may not take place, or it would take too long a time in view of the time that the country has already been in the ES period. If policymakers are not willing to wait any longer for the free market forces to bring about development, then they might ensure and accelerate the movement of economic development to the next stage, namely the medium stage.

An economy in the medium stage (MS) undergoes industrialization of the domestic market via import substitution. To a large extent the economy still remains dependent on the exporting sector developed in the previous period. The major impetus for changing the economy at this stage comes from the manufacturing sector, which enhances the size of the monetary market-economy. It also increases the country's productive capacity to supply the domestic market with domestically produced goods.

If the potential market is large enough, the level of industrialization may become fairly complete, up to the 'basic' capital goods and even machine-tool industry level, after import substitution in the consumer goods sector has been completed. The rate of growth of income increases with the employment of a greater number of people in the monetary market-economy, creating a wider domestic market for the domestically produced goods. The domestic manufacturing industry and the exporting sector use technology largely copied from a more developed country. The institutional framework of the financial system of the private sector is expanded over the previous one to supply domestic producers and traders with short-term loans.

The non-monetary subsistence sector decreases and the monetary market-economy expands to include the domestic manufacturing industrial sector. However, this stage of economic development is bound to a limit determined by the size of the domestic monetary market-economy.

An economy in the high stage (HS) of development undergoes a post import substitution phase in which agriculture and exporting manufacturing industries provide the major impetus of development. Modernization of the agricultural sector stimulates production of agricultural goods for domestic consumption as well as production of new 'non-traditional' goods for exports. The manufacturing sector also produces for domestic and foreign markets; as the rate of exports of these manufactured goods and 'non-traditional' agricultural goods increases, it generates an impetus in these sectors of the economy.

There is greater intersectoral trade. The manufacturing sector supplies inputs and technology to increase the productivity of the agricultural sector. In turn the agricultural sector supplies raw materials to the manufacturing industry and foodstuffs to the monetary market-economy's population. The size of the domestic monetary market-economy increases. It includes the urban, exporting manufacturing and growing part of the agricultural sector, namely that part that is undergoing a process of modernization in techniques of production to supply the domestic and foreign markets.

As the monetary base increases, the market size increases, and the monetary market-economy increases. During this period the subsistence sector of the economy disappears, the financial institutional framework expands, and there is a private money market where medium-term loans are provided to producers and traders. The state of the technological level is one of adaptation to domestic requirements of techniques developed elsewhere, and also of creation of endogenous technology that will reflect domestic requirements.

1. Policies

A successful movement to and completion of the MS period of development requires a radical change in the set of economic policies. These policies will be aimed at loosening some of the organic links between the domestic economy and the rest of the economies of the world.[5] The government will adopt policies to foster the growth of a domestic manufacturing sector and an appropriate financial institutional framework. The manufacturing sector enhances the monetary market and diversifies the domestic productive capacity. However, the success of this sector will require the creation of a financial institutional framework to foster the industrialization process. Thus, the government will establish general development banks and other financial institutions. This will provide the financial institutional framework that allows entrepreneurs access to long-term loans for long-term maturing investment projects that induce economic development.

Government will also pursue a discretionary policy by protecting and fostering the development of the sector from which the impetus will come, i.e. the government will pursue a protective policy toward the domestic manufacturing sector. Protection of the nascent manufacturing sector is the crux for the movement into the MS period. Protection will attract and encourage entrepreneurs to enter the various ventures of the manufacturing industry. This protection should last until the completion of import substitution of capital goods industry, where such industry is feasible. Once this has been achieved a new set of economic policies should be implemented.

To move into the high stage, a government which wants to push development will implement a set of policies aimed at fostering the creation of a 'non-traditional' export sector. 'Non-traditional' exports will come from the domestic manufacturing sector, as well as from other sectors whose output was not exported in previous stages. This sector will expand the domestic monetary market-economy and also increase the interchange of the other sectors of the domestic productive capacity. The set of policies will include protection and stimuli to the 'new exports' sector. The government may pursue some degree of reduction in the level of protection where appropriate in order to attain a degree of international competitiveness for the exporting manufacturing sector. This will increase international commerce, which in turn will stimulate 'new' exports. However, movement towards the advanced stage (AS) will require still another set of economic policies. These policies, like the former ones, will be implemented by the government responding to the interests of groups with enough political and/or economic power.

2. Politics

The movement to the MS period can be pushed by the segment of the population that does not receive or is not satisfied with the benefits of the early stage economic conditions. This group, in a democratic system, exerts a political pressure on the government to change the economic policies in order to extend economic benefits to them. This subset of the population is made up of those individuals who were not absorbed by the modern exporting or the traditional non-monetary subsistence sectors of the economy; they might participate in any or both of these sectors but, they are dissatisfied with the status quo. They will organize themselves into a political party (X) to exert political pressure over the existing government.

Assuming that the social and political institutions that serve the policies of the early stage are unchanged, the capacity of exerting any political influence over the government by the Party X, regardless of their political doctrine, will depend on their numerical and economic strength. Since they have organized themselves into a party to improve their economic position, it might be concluded that their economic influence is not considerable, thus their main influence will come from the numerical votes that they might cast in order to affect the outcome of the electoral process. If the political party is small, then the influence over the government will be ineffective. If the party is large, then they might influence the government's policies through their votes which would be reflected in the share of power in the government itself. However, the final outcome of the economic policies would be the result of the political compromises and possible alliances between the existing political parties and the new one(s).

If the political parties are of comparable strength then the economic policies that are the result of a political compromise will not have the necessary consistency or the required continuity to move the development process forward. In this case, in order to satisfy some groups, the government might foster the development of some sectors, giving the resemblance of movement to the next stage of development; however, the completion of the movement might take more time than some groups would be willing to accept, or it might never be completed at all. It is a necessary condition that economic policies of development be consistent in their design, implementation and continuity to complete a given stage of development. Finally, it is required that government policies should be flexible enough to allow for a change when the next stage of development comes about.

If the social and political institutions of the ES period are politically weak or destroyed then the emerging party might influence the government's policies. If Party X acquires control of the government, as many political parties try to do, it will attempt to perpetuate itself in power by implementing policies such that its constituency is benefitted and continues voting for it. Also, in order to justify its permanency in the government, the party will develop an all-embracing ideology, such that its policies are perceived by its citizens as to benefit the entire nation, not just a segment of it; thus, a nationalistic ideology will be the rationale used by a party or institution that attempts to implement economic policies that will move the process of economic development into higher stages. Nationalism allows the government to play an active role in the economy. Government expenditures on education, health, transport and some strategic manufacturing industries are perceived as a benefit to the entire nation, not just to one privileged sector. Also, it allows consumers easier acceptance of increases in the short-run costs resulting from the import substitution industrialization, and acceptance of some loss in terms of the neoclassical static 'efficiency' of the comparative advantage that the economy presumably has during the ES period for long-run benefits resulting from a diversified economy as the process of development moves to higher stages. These benefits will be reflected in terms of a higher standard of living, income, employment and relative economic independence resulting from a vastly increased domestic productive capacity. A relatively insulated economy will be shielded from the changes in the foreign demand for exportable goods or changes in the supply of importable goods that transmit the effects of the business cycle of the trading partners; thus, the domestic policy makers will aim at attaining development at a desirable rate of growth with less constraint than an open dependent economy.

THE EXPERIENCE OF BRAZIL

The argument will now be applied to Brazil, which is referred to as one of the Newly Industrialized Countries in Latin America. This new status was achieved in a relatively short time. Its experience will illustrate the argument.

Brazil maintained its characteristics of an economy in the ES period of development for nearly four centuries. The major export good changed over time, as a new good replaced an old one. These were essentially primary goods such as Brazil wood, sugar cane, minerals and coffee. These exports gave rise to a monetary market-economy that concentrated around the cities and exporting areas, and it coexisted with the traditional economy. Some of these goods were produced by slaves. When slavery was eliminated at the end of the empire in the late 1880s, an important change in the social and productive institutions took place. The power of the slave owners was diminished and some of the former slaves fled to the expanding south to join the labor force. The power of the Republic's government alternated, in a gentleman's agreement, between Minas Gerais and

São Paolo. These governments' economic policies were essentially that of free traders. The spurs of domestic production of importable goods during World War I were considered anomalous; thus, they paid only lip service to the industrialization effort of the Brazilian entrepreneurs.

The effects of the world depression finally convinced the major political leaders of the limitations of the Brazilian economy. The revolution of 1930 which brought to power President Getulio Vargas marked the definitive adoption of an import substitution industrialization policy. Import substitution industrialization added a new sector to the economy, generating an impetus to the development process and characterized the economy as one at the medium stage of development.

The domestic monetary market-economy that was once satisfied with imported goods was increasingly supplied with domestically manufactured ones. Government attracted and encouraged investment in the manufacturing sector by giving protection to domestic industry and providing long-term finance. Import competing industries received high rates of tariff protection, embargoes of competing imports (Law of Similars) and access to the preferential exchange rate. Long-term loans were provided by the 'Industrial Window' and subsequently by the National Development Bank and other government-owned financial institutions to those industries that were regarded as desirable for Brazilian development. Finally, the government itself invested in those sectors that the leaders considered strategic to the national interest such as the steel industry (Volta Redonda) and oil industry (Petrobras). As a consequence, import substitution of consumption goods, such as textiles, was completed in the 1940s. Hence, they began substituting imports of intermediate and capital goods. Brazil completed the substitution of capital goods by the early 1960s.

The thrust of import substitution industrialization had various effects. It diversified the Brazilian productive capacity, enhanced the level of domestic employment and income, incorporated a greater number of people in the monetary market-economy, and decreased the size of the non-market economy. The purchasing power of the market economy was at least 6,000 dollars in 1980 and comprised of at least five per cent of the population.

With the addition of the exporting manufacturing industries and modernization of agriculture, Brazil entered into the HS period in the mid 1960s. The government implemented a set of economic policies aimed at stimulating the development of 'non-traditional' exports coming from the manufacturing and agricultural sectors. This generated a new spurt in the developing process.

The manufacturing sector then produced for the foreign as well as domestic markets. Exports of this sector were consumption goods such as shoes, textiles and furniture and durable goods such as cars, buses and electrical, mechanical and military equipment. Part of the agricultural sector took advantage of the impetus and technology supplied by the manufacturing sector. And, with the aid of the government, modernized and diversified its production for domestic consumption and exports of non-traditional goods such as soya beans. As the rate of these new exports increased there was an impetus in economic activity in which these two sectors played an important role.

The modernization of the agricultural sector had the effect of intensifying or introducing the monetary market-economy into areas of the North, East and South not previously included. The non-market economy remained largely in the West and in isolated pockets in the other regions. These pockets will disappear as the economy reaches the advanced stage of development.

Getulio Vargas and subsequent military administrations ruled Brazil with a nationalistic ideology. The São Paolo–Minas Gerais power was neutralized by President Vargas who came to power with the help of the military, proclaiming to be an alternative to the status quo. During his administration and the so-called Vargas system that remained in place after him, workers as well as the emerging entrepreneurial class organized themselves into two political parties: the Brazilian Workers Party (PTB) and the Social Democratic Party (PSD). These parties supported Vargas during his first administration, his re-election, and the subsequent elections of the followers of his establishment such as Juscelino Kubischek.

The nationalistic ideology, though with some slight differences in the nationalism concept among its leaders, basically maintained that what the Vargas system was doing was good for Brazil. It permeated various strata of Brazilian society. The emerging entrepreneurial class, urban workers and a segment of the military (the *tenentes*) supported it. During the late 1940s popular demonstrations under the banner '*O petroleo e nosso*', (the oil is ours), forced the government to take an active role in the petroleum industry and barred foreign investment in some parts of this industry. As a result the government entity Petrobras was created. Petrobras is active not only in the oil industry, but also in the petroleum related industries such as petrochemicals.

The 'endogenizing' of technology is another field in which the government has stimulated development, although private firms have contributed to this. After the turn of the century, most Brazilian firms used the technology of the country where their equipment came from. There was not a conscious effort, nor were expenditures made, to adjust it to the Brazilian environment. This may well have been due to output technology rigidity, the levels of production in general being low for the small domestic market, so it was easier for firms simply to copy imported techniques of production. However, firms installed in the late 1950s and early 1960s did not merely copy foreign techniques of production any longer.

If the output technology hypothesis, which maintains that the choice of technique is governed by the scale of production, is true, then the larger Brazilian domestic market of those years required a higher level of output. As a consequence firms had some flexibility in adjusting their technology to the host country's environment. Moorley and Smith, commenting on technological adaptation of firms installed in the late 1950s and early 1960s, claim that there were 'substantial differences in production techniques between multi-national firms and their subsidiaries in Brazil,[6] however, the importance lies in the fact that the 'evidence suggests that adaptations by multi-nationals to local conditions are substantial'.[7] This seems to support our contention that at a HS period there is technological adaptation in production.

The transfer of foreign technological knowledge to Brazil occurred mainly via licensed and joint ventures. Brazilian entrepreneurs produced many manufactured goods under licence or by buying patents of production. For instance, the pharmaceutical industry had laboratories engaged in the research and development of new drugs in the 1950s. Furthermore, Decree No. 72522 of 1973 required that a percentage of gross sales be allocated to research projects to intensify endogenization of technological knowledge.

Technological knowledge was also transferred through joint ventures of private domestic and foreign capital as in the case of synthetic fibers used in the textile industry, and through the tri-pe (three leg: state–domestic private–foreign) joint ventures as in the case of petrochemical industry.

With the installation of the capital goods industry, Brazil had a more complex industrial productive capacity with a higher level of technological sophistication than she did during the early phases of the MS period. The Brazilian authorities were not content to have only adaptation of the know-how to the local conditions. They wanted to endogenize the technological 'know-why', so they made a considerable effort at allocating resources to research and development.

For the purpose of creating a Brazilian technology, the First National Development Plan (I PND) called for the government to allocate funds to technological development at a rate of 100 million cruzeiros a year after 1968. However, when the Basic Plan of Science and Technological Development (PBDCT) was approved, that sum was increased to 4.3 billion cruzeiros (717 million dollars) for the 1973–1974 period. The task of the plan was of 'putting modern science and technology at the service of Brazilian Society in its fundamental objectives.'[8] It sought to obtain modern technology for the priority sectors and gradually to produce domestic technology.

The master aspects of the PBDCT include development of new technology in nuclear energy and space research applied to development; oceanography and information systems (computers) as well as in high technology and capital intensive industries such as electronics, aerospace, chemistry and precision mechanics; strengthening of the technological absorptive capacity of national enterprises, both public and private; strengthening of research organizations in priority sectors and expansion of the supply of trained researchers; consolidation of the support system through the establishment of a national scientific information system; integration of university, business and government in research, with realistic problems and objectives. The Brazilian government's objectives include making the country an exporter of computers, locomotives and generators.

The effort to develop a Brazilian technology at the HS period is a conscious attempt to break away from the foreign technological dependency. It also seems to indicate that Brazilian authorities are attempting to substitute a Brazilian produced technology for a foreign one. The beginning of an endogenous technology contradicts dependency theorists' claim that a developing country will always be dependent technologically.

Brazilian technology will reflect the scarcity and needs of the domestic economy. Thus, the technological innovation of an alcohol-powered Brazilian car reflects the relative gasoline scarcity and alcohol abundance. After the oil shock of the 1970s the government stimulated further the production of the alcohol-powered car.

The entrepreneurs supported the system through PSD, which was especially active during Kubischek's administration. This group successfully opposed the implementation of the IMF designed economic policies; they wanted to have an uninterrupted access to the government credit lines as well as tariff protection for their products. The military also supported nationalism, while the *tenentes* who had supported Vargas became senior officers in the 1960s. When the policies of João Goulart's administration slowed down the rate of economic growth and the development process, General Castello Branco with the help of other former *tenentes*, such as Eduardo Gomez, Magahlanes, etc., took over the government and reactivated the process of development.[9] Under the military's tight rule nationalism as an ideology was still emphasized.

The economy grew at an average rate of 10 per cent per annum. This period is known as the 'Brazilian Miracle'. This high rate of growth was achieved in a relatively insulated economy; however, Brazil accumulated a 105 billion dollar foreign debt in the process. Brazil's exports value was 21.8 billion dollars in 1983, of which 13 billion dollars – 59.7 per cent of exports – were industrial products; more specifically, manufactured products made up 51.6 per cent, while coffee exports, which used to be the main export good in the ES period, accounted for only 9.5 per cent of export value. This illustrates that diversified exports came from a diversified productive capacity in the Brazilian economy.

CONCLUSIONS

It has been emphasized that the government used nationalism as an ideology and rationale for its policies. The political and economic conditions under which development takes place have been outlined, and the rudiments of appropriate government policies have been presented. The model has been examined in the light of the experience of Brazil, a country which has reached the higher stages of economic development. The policies that were implemented have been outlined through which Brazil has initiated its technological endogenization. Economic development in a mixed economy, therefore, in the latter part of the twentieth century requires that the government adopt appropriate and consistent economic policies to move the development process into higher stages.

Notes

1. W. W. Rostow, *The Stages of Growth* (Cambridge: Cambridge University Press, 1971), p. 150.
2. K. Marx and F. Engels, *The Communist Manifesto*, ed. by S. H. Beer (New York: Appleton-Century-Crofts, 1953).
3. G. Frank, 'The development of underdevelopment', in R. I. Rhodes (ed.), *Imperialism and Underdevelopment* (New York: Monthly Review Press, 1970).
4. E. Carreño Ramón 'The Brazilian model of development', unpublished dissertation.
5. M. A. Heilperin, *Studies in Economic Nationalism* (Paris; L'Institut Universitaire de Hautes Études Internationales, 1960), p. 27.
6. S. A. Moorley and G. Smith, 'The choice of technology: multinational firms in Brazil', *Economic Development and Cultural Change*, Vol. 25 (January 1977), pp. 260–61.
7. Ibid., p. 246.
8. 'Plano basico de desenvolvimento cientifico e technologico do Brasil', *Conjuntura Economica*, Vol. 28 (January 1974), pp. 48–51.
9. R. J. Alexander, 'The Brazilian tenentes after the revolution of 1930', *Journal of Interamerican Studies and World Affairs*, May 1973.